THE PERIOD LITERACY HANDBOOK

Everything you need to know about your menstrual cycles, period.

Dr. Anne Hussain, ND

Library and Archives Canada Cataloguing in Publication

Hussain, Anne, author

The Period Literacy Handbook / Dr. Anne Hussain, ND

ISBN: 978-1-7390104-0-9 (paperback), 978-1-7390104-1-6 (ebook)

Cover art by Katerina Naskovski
Drawings by Saira Hussain
Diagrams by Anne Hussain
Published in Canada

For more information, email hello@annehussain.com

For Ammi & Abbu

CONTENTS

INTRODUCTION

The average person with periods goes through more than four hundred menstrual cycles in their lifetime–over four hundred cycles of changing hormones that impact mood, energy levels, productivity, appetite, metabolism, sleep, connection, and more. Yet we don't learn enough about periods. In fact, in some parts of the world, we don't learn anything at all. We study, on repeat, the first twenty elements on the Periodic Table and how to calculate the acceleration of a train from rest–which are important–but not as important as learning about our bodies.

This book is all about the menstruating body. It's for anyone with periods or who wants to learn about menstruation. It contains answers to questions like:

- What happens in a menstrual cycle?
- Is my period "normal"? How do I tell if something is off?
- Do I need to get my hormones tested?
- Did I cause my period problems?
- Should I quit high-intensity exercise for my hormonal health?
- Which foods, if any, will make my periods better?
- Why are menstrual and hormone-based conditions dismissed so often?
- How do I talk to my doctor about my period problems?
- What kind of contraception is best for me?
- Should I sync my life with my menstrual cycle?
- What is my life, even?! (I kid, I don't have the answer to that last one.)

These are a small sampling of questions I've kept in mind while writing this book.

You see, this is the book I wish I had growing up.

I remember when my first period started. It was 2001 (... the good ole post-Y2K days when *Hanging by a Moment* by Lifehouse was the song of the year and the first *Harry Potter* and *LOTR* films came out). I was in Karachi, Pakistan, where I lived between the ages of five and thirteen. I came home from school one day only to realize I was bleeding ... from the inside. As my physical body froze from shock, my mind was running a mile a minute.

There's something wrong here. I don't think I hurt myself. Did I hurt myself? Did I eat beets? Should I call for Ammi (my Mum)? What will happen to my family if I die?

As sensation returned to my body, I started pressing my abdomen, pelvis, vulva, and back to determine where the heck this blood was coming from. I searched in vain; the blood was coming from the *inside*.

Heart pounding in my chest, I gingerly pulled up my pants, washed my hands, and went to find Ammi. I told her that there was something wrong with me. Tears welled up in my eyes as I told her that I was peeing blood.

To my surprise, she did not break down in sobs. Instead, she chuckled, made a comment about how she thought I had more time, and proceeded to tell me that this happens to young girls. She got me a pad, instructed me on how to use it, told me that I'd be bleeding for a few days, and said it would happen approximately monthly.

That was it–that was my menstrual education. That's the information my mum had received growing up, and that's what she passed on to me.

I was thirteen years old, and I had never heard of periods. No one in my life talked about them (at least openly). I had learnt nothing in school about reproductive health because, well, it was not part of the curriculum in Karachi. I had been taught over and over again that William Shakespeare lived between the years 1556 and 1616 (a fact I still remember to this day), but I learnt zilch about a process that I experience every month.

When I think back to that time, I remember watching pad commercials on TV– albeit for a brief moment before the channel was changed–having no idea what those commercials with the blue liquid were about. Then, as I did realize what a period was, I adopted the hush-hush culture around periods through osmosis, never speaking to anyone about it apart from Ammi whenever I needed my stash of pads replenished, or the odd friend to check for leakage.

When my family moved back to Canada in time for me to start high school, I still hadn't learned about the menstrual cycle, as it's taught in elementary school here. I did have a sex ed class, where most of what I learnt was about abstinence, sexually transmitted infections, bananas, and condoms, but not so much about periods.

Throughout high school, my cycles were quite irregular, sometimes pulling a disappearing act for months on end. I have a mild, genetic kind of anaemia called beta-thalassaemia (common in people of African, Mediterranean, and South Asian descent), so we chalked up all my unrelenting fatigue, irregular cycles, and varying flow to my youth and that for a bit. Then, things started to change.

As I started at McMaster University in my birth town of Hamilton, Ontario, not only did my cycles become more irregular, but I also started to break out in acne, have low mood, and experience mild hair loss. I went back to my family doctor, and she offered the solution presented to most people with periods and hormonal problems (and often without further investigation or discussion): the birth control pill.

What ensued was a downward spiral to feeling worse than ever before followed by poring over research studies and textbooks, coming off the pill, conversations with my family doctor and a naturopathic doctor, a slow journey to getting a polycystic ovarian syndrome (PCOS) diagnosis, regulating my cycles with some help, becoming a naturopathic doctor and menstrual health advocate, and helping others make educated decisions for their health–a journey, I'm happy to share, I'm still on to this day.

Now, I get to support others with their periods, hormones, and fertility; PCOS patients make up the largest portion of my practice. My biggest job is to listen to my patients' stories and experiences because that's what allows me to assess them comprehensively, run appropriate testing, understand their goals, work with and refer to other members of their healthcare team, teach them about their bodies, discuss treatment options (including meds) so they can make informed decisions, and revisit their health plans to reflect changes in their health and life (and science!). I don't always get things right (who does?!), but I strive to do the best I can by keeping up to date with research, learning from my patients and teachers, and working on my biases.

I have grown up in a culture and time where menstruation, menstrual research, and menstrual talk have been women's domain even though it is not only women who menstruate or are affected by menstruation. I've tried, as much as possible, to be inclusive in my language. Research is already becoming more nuanced and inclusive in its scope, so hopefully, we'll have a wider breadth of research pertaining to more diverse groups soon.

With that acknowledgment, I must also address some of the privileges I have: I am a cis-gendered able-bodied heterosexual female who has had the privilege of living in two different countries, currently living and practicing on the traditional territory and home to the Anishinaabe, the Chippewas, the Wendat, and the Haudenosaunee peoples. I have freedom of speech, easy access to period products, professional-level education, and access to a public healthcare system (even though that system is imperfect, sometimes looks down on my profession, and is under threat by our current conservative provincial government) amongst other resources.

What I know is the product of what I have learnt with the help and work of many teachers, mentors, researchers, scientists, doctors, advocates, patients, friends, family, online forums, and different populations who were willing and unwilling to be involved in research and experimentation.

This book wouldn't be before you in the way it is without this work, knowledge, support, and experience. Honestly, it would probably look different if it weren't for the socio- and geopolitical conflicts that the early 2020s have brought with them. I have worked and reworked these pages many times over, trying to capture as much nuance as I can.

I've probably missed things, overused phrases and sentences, been cavalier with my words (unintentionally), over-simplified or over-explained concepts, gone on tangents, overlooked some of my biases and blind spots (I am human, after all), and presented information that will likely become outdated. I'm sure future Anne would write a slightly different book—such is the nature of things.

I came up with the name of this book before I even had a real plan. In my mind, period literacy referred to having a basic understanding of the menstrual cycle, cultivating an awareness of it, and holding all of that in the context of your life and the world we live in. These concepts are not new and have been around for a long time, with different names around the globe. In fact, as I was writing this book, the Terminology Action Group of the Global Menstrual Collective formalized the definition of *menstrual health*, which is "a state of complete physical, mental, and social well-being and not merely the absence of disease or infirmity, in relation to the menstrual cycle."[1] I love this term. Its full scope includes age-appropriate information, hygiene practices, accessibility, healthcare, respect without stigma, and unhindered participation in life. Add to that some socio-political and cultural context, and that's what this book you're currently reading or listening to is about.

There are thirty-two chapters. This, coincidentally, happens to be the average length of my menstrual cycle (... or, perhaps, that was a subconscious manifestation!).

This book is *not* a how-to guide to fix your periods in thirty days or an offering of "biohacking" tips for your menstrual cycle. It is, however, full of the basics of hormone health with added nuance that we sometimes forget about in this fast-paced age of instant gratification, short-form content, and TLDR (too long, didn't read) captions. It contains evidence-based strategies with a dash of cultural and traditional wisdom as well as a smattering of socio-political commentary and personal opinion for you to consider in your menstrual journey. It is not medical advice but for educational purposes only.

---

[1] *Terminology Action Group of the Global Menstrual Collective, 2019*

Part One is all about the science of menstruation. You'll learn (or re-learn!) about the communication system between your brain and reproductive organs that lead to a menstrual cycle. You'll get reacquainted with specific language pertaining to reproductive anatomy and hormones, all the while understanding how periods change over time, what a healthy menstrual cycle can look like, and why the often-demonized estrogen is really important for your health. This part reads kind of like a casual textbook; it even has anatomical illustrations and diagrams.

Part Two covers cycle tracking: how to best track your cycle and what to track in the first place. This will help you notice what's happening in your body, understand your personal menstrual patterns, and recognize when something is off. This is one way to start listening to your body and use the information gathered to understand what it's telling you.

We take a bit of a turn in Part Three and dive into contraception, from hormonal methods to fertility awareness ones. The pill and IUD, which are commonly used globally, are discussed in detail so you can make decisions with your healthcare team in an informed manner. Too often, the pros and cons are not adequately considered when these methods are prescribed, even though these interventions are used not only for contraception but also for a plethora of other hormonal concerns like acne and pain. The chapters in this part of the book will hopefully offer some food for thought to facilitate your decision-making around contraception.

Part Four touches on themes in general menstrual health as well as common menstrual conditions like PMS, painful periods, PCOS, and endometriosis. I haven't covered *all* the possible menstrual and hormonal concerns but picked the ones with a high global prevalence. Whether you suspect a diagnosis, already have one, or have non-specific symptoms of unknown origin, this part of the book is worth a read. It lays out diagnostic criteria, symptoms, testing, and treatment options (both conventional and naturopathic). Menstrual conditions can lead to missing out on work, school, and life, and too often they are normalized or dismissed. I encourage you to go through this section especially if you feel like something is off; you might just find your experience reflected in one of the diagnoses.

Part Five is a broader discussion surrounding period health and how your menstrual experience is multifaceted—it is impacted by your micro- and macro-environments. Social and structural factors like geography, education, policy, patriarchy, the healthcare system, capitalism, and the world at large all affect our lives and health. Period products and menstrual activism are also covered in this part. These chapters will hopefully help you understand why so many period problems are normalized, what's under your control and what's not when it comes to your periods, and how you can move your menstrual health forward as an individual and as part of the human collective.

And last, but certainly not least, Part Six brings together all the concepts in this book and offers real-life guidance for hormone health. There is something for everyone in this section. Nutrition, exercise, sleep, supplements, testing, and self-advocacy tools are abundant in the final few chapters so you can start working toward better menstrual health today. I review some trends and info popularized by TikTok and Instagram, highlighting the many benefits of the internet while doing some good ole myth-busting so that you're spending your precious effort, money, and time on the right things. The concept of leaning into your cyclical framework, following the natural ebb and flow of hormones, is also covered. This part of the book is probably what most of you are looking for; it contains actual actionable steps for more peaceful periods.

You can read this book in order, or simply skip to the chapters or parts you're interested in. To keep it digestible, there are infographics, illustrations, chapter summaries, and a glossary at the end. I hope it will serve as a resource that you can turn to time and time again.

I've written this book as a vehicle of education, awareness, self-advocacy, and food for thought as you move through your health and life.

You don't have to love your periods (honestly, I can't say that I do), but a basic understanding allows us all to come to terms with this fact of life ... and, perhaps, have a better relationship with it (bonus!). That's what I've been working toward in my own life.

So, here's my offering to thirteen-year-old Anne, whose first period took her by surprise. Here's my contribution to patients who are looking to further their understanding of their hormones as most of health happens outside of the doctor's office. Here's my contribution to the people in my life who come to me with great questions that I often give very long-winded answers to. Last, but certainly not least, here's my contribution to you, dear reader or listener–I'm so glad you're here learning about your body.

Our views on some of the topics covered might differ, but I hope that you find something to engage with, something to think about, and, perhaps, something to look into more deeply.

Health and healing are not finite goals, nor do they follow a linear path. What works for someone else might not work for you. Managing a complex condition will look different than prevention and health promotion. Heck, what health means to you is probably different from someone else. As you navigate the next thirty-two chapters, I encourage you to reflect on that definition and see how you can move toward it.

PART ONE

MENSTRUAL ANATOMY & PHYSIOLOGY

CHAPTER 1

Menstrual Anatomy

The menstrual cycle is a combination of two cycles: the uterine cycle and the ovarian cycle; it describes what's happening at the level of the uterus as well as the ovaries. These organs aren't the only ones involved. In fact, the menstrual cycle is the result of many organ systems working collaboratively and communicating through your hormones (more on that later).

For now, let's start with an anatomy lesson. If you've used the words vulva and vagina interchangeably, or the catch-all term "down there", you're not alone.

Understanding (at least a little bit about) your menstrual anatomy is important so that you have a good grasp on where your organs are located, learn the differences that can exist from person to person, communicate more clearly with your healthcare providers, and be on the same page as we navigate this book. Plus, it might help you appreciate just how cool the body is!

Uterus

The primary organ related to menstruation is the uterus, which is found in the centre of the lower abdomen. It's located between the urinary bladder and rectum. It has three layers:

- perimetrium (the outermost layer)
- myometrium (the middle, muscular layer)
- endometrium (the innermost layer)

The endometrium–the layer that is the focus of this book–grows throughout the menstrual cycle and is sloughed at menstruation time. The endometrium itself has two layers, the basal layer and the functional layer; it's this functional layer that grows over the cycle and then flows during a period.

Similar to how external features can vary from person to person, our insides can also look different. The uterus can be found in different shapes and sizes. Typically, the inside of the uterus is one big cavity; however, you could have a partial or full divide inside. If you've ever looked at pelvic ultrasound results (or uterus diagrams online if you're anything like me), you might have seen the words anteverted, retroverted, anteflexed, and retroflexed. These terms are simply referring to the tilt at the base and top of the uterus. The most common orientation is anteverted and anteflexed, and that is what's depicted in Diagram 1.

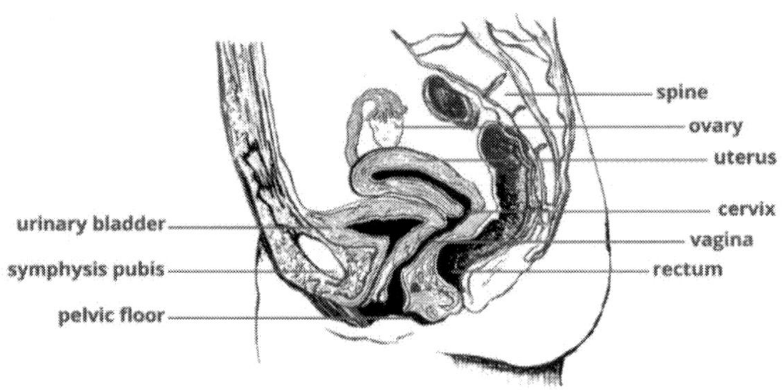

DIAGRAM 1: *Lower abdomen side view. The uterus sits in between the urinary bladder (in front) and rectum (behind). The vaginal canal extends from the bottom of the cervix. The spine can be seen behind the rectum, and the symphysis pubis (the joint between the left and right pubic bones) can be seen in front of the bladder. The pelvic floor is at the base of the organs but cannot be fully visualized.*

At the base of the uterus is the cervix, which is essentially a cylinder with an opening at each end. The cervix is technically part of the uterus, and it connects with the vaginal tract at the bottom. It has numerous glands that produce the clear or whitish mucus that you may see throughout your cycle, especially around the midpoint when ovulation occurs. The cervix is what dilates during labour for a baby to pass through to the vaginal canal and out into the world.

Unfortunately, it is also the organ with the third-most common cancer–cervical cancer–in females worldwide, and that cancer is almost always associated with human papillomavirus (HPV) infection. HPV can also cause cancers of the vagina, anus, and vulva. That is why regular screening through Pap smear, liquid-based cytology, or HPV DNA testing is important. These tests require a small sample of cells to be obtained from your cervix and sent to a lab for testing. The sample can be collected using a swab in your doctor's office or an at-home kit.

Fallopian Tubes & Ovaries

On each side of the top of the uterus are the fallopian tubes, which are also called oviducts or uterine tubes. The oviducts are kind of like a roadway for egg transport, for sperm to travel to the egg, and for the fertilized egg to travel to the uterus. They are hollow, commonly ten to twelve centimetres in length, and end in finger-like projections called fimbriae; it is these fimbriae that connect with the ovaries as seen in Diagram 2 along with the fallopian tubes and uterus.

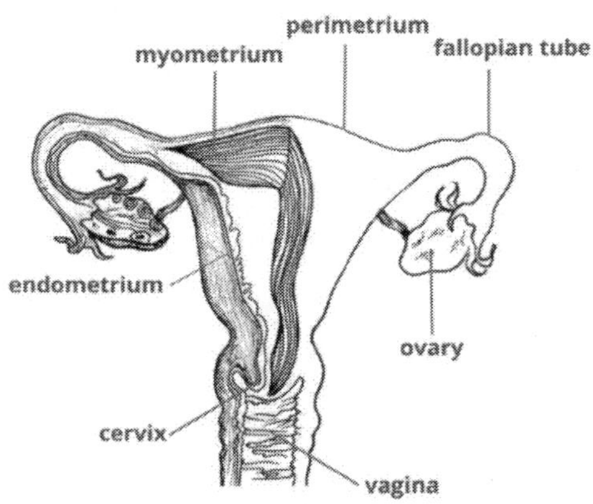

DIAGRAM 2: *Uterus front view, half peeled back for a view of the inside. The uterus has three layers: perimetrium (outer), myometrium (middle), and endometrium (inner). The fallopian tubes are on either side of the uterus, ending in fimbriae that make contact with the ovaries. The bottom part of the uterus is the cervix which opens into the vaginal tract. The inside of the vagina has folds called rugae.*

The ovaries are paired organs, with each being approximately the size of a golf ball. They are found in the lower left and lower right areas of the abdomen. Rarely, people may be born with only one ovary or more than two ovaries. The ovaries have two primary functions: to produce hormones like estrogen, progesterone, and testosterone, and to house your egg cells, which are called oocytes.

Vagina

At the bottom of the uterus, connected to the cervix, is the vagina, also referred to as the vaginal tract or vaginal canal. It is a fibromuscular tube that extends from the cervix to the external genitalia.

It is the exit route for menstrual blood and tissue from the uterus, is involved in sexual functions such as arousal and pleasure, has a role in reproduction for sperm to enter and a baby to exit, and offers immune defence.

It has an acidic pH of 3.8 to 4.5, secretes fluid, and has its own microbiome containing microorganisms like bacteria (kind of like how your gut has lots of microscopic dwellers). The bacteria that are most prevalent in the vaginal tract are *Lactobacillus* strains, which help support vaginal health and influence immune responses. Similar to the differences in gut bacteria from one person to another, we have differences in the vaginal microbiome as well. Maintaining all of these parameters is crucial for good vaginal health, which is why the use of soaps or cleansers internally is not recommended. These can strip away protective compounds, disrupt the pH, and/or disturb the balance of microorganisms in the vagina.

The inside of the canal has folds called rugae, which look kind of like the bunching at the bottom of leggings or tights that are slightly too long for you. The vagina opens into the external genitalia at the bottom, and this opening may be covered with a thin membrane called the hymen.

The hymen doesn't really serve a biological purpose as far as we know. It usually ruptures in the first few days of life, and you may or may not have some remnants of it as you grow older. Rarely, it'll stay intact, which is called an imperforate hymen; this requires medical attention since it blocks the vaginal opening completely. The hymen, two-finger, or per vaginal exam historically–and inaccurately–was (and sometimes even now is) used as a form of so-called virginity testing even in doctor-patient settings. If it was intact, you were classified as a virgin and, if not, well, then there were consequences. Virginity testing has been condemned by human rights and international health organizations as well as scientists and doctors, but it is still practiced in many countries and by many healthcare professionals. To be clear: there is no basis for so-called virginity testing.

External Genitalia

The vagina ends in a structure called the vestibule, which is a transition area between your insides and outsides. The external genitalia include the vulva, which is the part of your body that touches your underwear and is associated with those assigned female at birth visualized in Diagram 3.

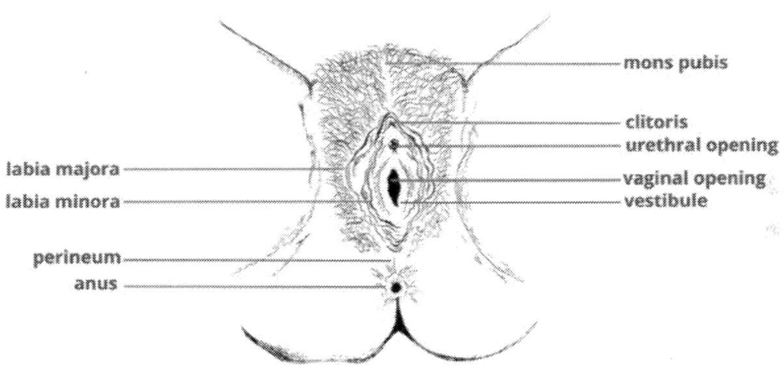

DIAGRAM 3: *Bottom view of the external genitalia. From top to bottom in this illustration is the mons pubis, the visible part of the clitoris, the urethral opening, the vestibule and vaginal opening, the perineum, and the anus. The labia minora and majora are also visible on each side of the central structures. Hair covers the mons pubis and the skin around the labia.*

The term vulva essentially describes all the structures of the external genitalia and includes:

- labia minora (smaller inner lips)

- labia majora (larger outer lips)
- clitoral glans and hood (external portions of the clitoris)
- mons pubis (fatty cushioning atop your pubic bone)
- urethral opening (also part of the urinary system since this is where urine exits)
- vaginal opening
- perineum (tissue at the bottom of the labia that ends at the anus)
- glands that are present around those structures

The vulva has roles in the sexual response, menstruation, and birthing process. It also offers mechanical and chemical defences against contaminants and pathogens.

I have many patients who share that their vulvas do not look like the ones they've seen online. And that's because vulvas come in all kinds of shapes and sizes. Genetics, hormones, age, background, surgical procedures, trauma, skin conditions, and other factors play a role in what your external genitalia look like. Symmetry, colouration, hair distribution, folds, length, and width can all vary from person to person. Just like there are differences in how our noses and ears look, there are variations in how our genitalia look, and this is totally okay!

BOX 1. What's that smell?

The vulva has glands that produce pheromones, sweat, and sebum that can be broken down into odorous compounds by skin bacteria. This means that there can be a smell associated with the external genitalia. Combined with secretions from the cervix and vaginal tract, especially when cervical mucus is being produced mid-cycle, a distinct odour can be present. This is normal.

It does not mean you need to use harsh soap or antimicrobial wipes. It also does not mean that soaps and perfumes should be used internally. Plain ole water is what works best! To cleanse the hair around the vulva, a gentle cleanser that is slightly acidic can be used so as to not disrupt the protective substances, fatty acids that reside on the skin, or the pH of the external genitalia.

The vaginal tract has a pH of 3.8 to 4.5 and the vulva has a pH of 5.3 to 5.6. Staying within these ranges allows these parts to remain healthy. Given the proximity to the anus, this area is designed to handle microscopic levels of feces and other compounds.

An infection can cause itching, burning, foul odour, and abnormal discharge. If you're noticing any of these, or have questions about your hygiene or vulvar health, please address this with your healthcare team.

Pelvic Floor

Last in this chapter, but not least, we'll cover the pelvic floor, which is the group of muscles to which the pelvic organs (uterus, urinary bladder, and rectum) are anchored. These muscles are found between the tailbone at the back and the pubic bone at the front. The pelvic floor includes muscles (like the coccygeus and levator ani) and the pelvic diaphragm, which is like a hammock for your pelvic organs.

Although not technically a reproductive organ, the pelvic floor plays an important structural role in your anatomy and can be involved in period pain, chronic pelvic pain, endometriosis, incontinence, surgical scars and adhesions, and prolapse of the rectum, bladder, or uterus.

Kegels are not the only way to address the pelvic floor although they can be part of a strategy where muscle tone needs to be increased! The coordinated response of these muscles is what is often key and will not be fully addressed by Kegels alone. Pelvic physiotherapists can assist with general pelvic floor health and any issues pertaining to it, especially if you have any of the concerns listed here. Lower back pain, muscle tension in the legs, and pregnancy-related musculoskeletal concerns can also benefit from pelvic floor work.

CHAPTER HIGHLIGHTS

- The main organ involved in menstruation is the uterus, which has three layers. The lining of the innermost layer (the endometrium) grows during a menstrual cycle and is shed during a period.
- The cervix is found at the base of the uterus and has glands that produce mucus.
- The ovaries are where sex hormones are made and also where egg cells, called oocytes, are housed. The ovaries are connected to the uterus via fallopian tubes or oviducts.
- The vagina is a muscular tube found on the inside of the body. It runs from the cervix to the vulva. It has an acidic pH and houses microorganisms, and this balance can be disrupted if soaps or cleansers are used internally.
- The vulva refers to structures found on the outside of the body including the labia majora, labia minora, mons pubis, clitoris, urethral opening, vaginal opening, vestibule, perineum, glands, and pubic hair. Plain ole water works best for cleansing to not disrupt the skin barrier and pH.
- Vaginal discharge and a mild vulvar odour are normal. These can change slightly depending on where you are in your menstrual cycle. Symptoms of burning, itching, suspicious discharge, and/or foul odour should be discussed with a healthcare provider.

CHAPTER 2

Menstrual Cycle Overview

The menstrual cycle results from a fascinating and complex hormonal messaging system between the brain and ovaries. This network is called the hypothalamic-pituitary-ovarian (HPO) axis, which is activated at menarche, i.e., your first period ever during puberty. The signals from the brain lead to responses in the ovaries and uterus cyclically, resulting in what we know as the menstrual cycle.

The menstrual cycle is an amalgamation of two cycles: the ovarian cycle and the uterine cycle. Put more simply, it represents the concurrent changes that are happening at the levels of the ovaries and the uterus.

Ovarian Cycle

This is the series of events that occurs in the ovaries and consists of:

- The follicular phase, which is when egg-containing follicles grow.
- Ovulation, which is when an egg is released from a follicle (called the dominant follicle).
- The luteal phase, which is when the dominant follicle's empty shell secretes progesterone as a result of ovulation.

Uterine Cycle

This is the series of events that occurs in the uterus and consists of:

- Menstruation, which is when the endometrial lining of the uterus is shed, i.e., your period.

- The proliferative phase, which is when the endometrial lining grows thanks to estrogen.
- The secretory phase, which is when progesterone stabilizes and matures the endometrial lining.

Menstrual Cycle

Layer together the events and terminologies of the ovarian and uterine cycles, and voilà! We have the menstrual cycle.

It technically consists of two halves: from menstruation to ovulation is the follicular phase, and from ovulation to the next period is the luteal phase. However, it is usually divided into four parts to create more separation between the distinct hormonal and physiological processes that are occurring. This subdivision makes the cycle a bit easier to study and understand. The four parts are as follows:

1. **Menstruation**, aka your period, the event during which you're shedding the endometrial lining of the uterus.

2. **Follicular phase**, which is the timeframe when your egg-containing follicles as well as endometrial lining are growing. One of the follicles becomes the dominant follicle.

3. **Ovulation**, the short-lived event when the dominant follicle releases an egg (or, an oocyte).

4. **Luteal phase**, which is the stretch of time from ovulation to your next period. The shell of the dominant follicle makes progesterone to stabilize the endometrial lining at the beginning of this phase. This is followed by a drop in hormones toward the end of the phase as the body realizes that pregnancy didn't occur.

And then, we go back to the beginning! Diagram 4 is a depiction of the two halves of the menstrual cycle. There's a lot more that happens, and we'll get to that in Chapter 3, but let's get into how long this process takes first.

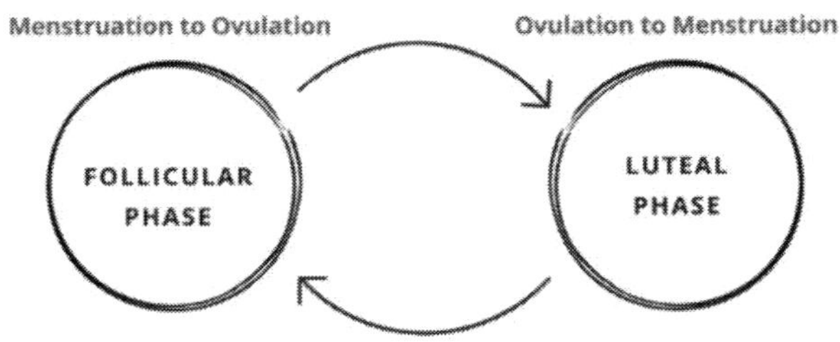

DIAGRAM 4: *The two halves of the menstrual cycle, (1) follicular phase from menstruation to ovulation, and (2) luteal phase from ovulation to menstruation.*

Cycle Length

The *average* length of a menstrual cycle–from the beginning of one period to the next period–is approximately twenty-eight days. It does *not* have to be exactly twenty-eight days, and it's not something you need to strive for. A healthy menstrual cycle that has been established can be anywhere from twenty-one to thirty-five days.

Cycle Day One is the day you experience full flow. It can take hours for flow to build up, which is normal. If you experience spotting for a few days before your period, the spotting is considered the end of your previous cycle, and the new cycle begins once your bleeding surpasses spotting into a more sizeable volume.

To calculate your cycle length, you're going to start with day one of your period bleed and count until the day before your next period starts. For example, if your period starts on May 1st and the next one starts on May 30th, your cycle length for May is twenty-nine days. If you use an app, you should be able to find or calculate your average cycle length once you've had a few periods tracked as well as a breakdown per individual cycle.

The length of your cycle should be fairly uniform for you, whether that stretch is twenty-five days or thirty-two days. If it's off by a few days on either end, it's still considered a regular cycle. Ovulation is what will determine the regularity of a cycle as you'll discover in Chapter 3. We're talking about a natural cycle that is unmedicated here. Chapters 9 and 10 contain details on bleeding patterns while on hormonal contraceptives. You might have the odd time when your cycle is longer or shorter than usual. This is totally normal! The body will respond and adapt to the goings-on in life and may delay or skip ovulation temporarily, for example, if you've been ill or have had a huge amount of stress. Typically, everything will return to your normal rhythm within those circumstances passing.

There are also times when we expect variability in cycle length—short and/or long—like in the first year of periods starting up, in the lead-up to menopause, and during lactation. This is due to the hormonal fluctuations and interference in ovulation inherent to those transitions. More on that in Chapter 5.

Irregular cycles are medically defined based on where in your menstrual cycle journey you are, for example:

- Irregular cycles are less than twenty-one days or greater than forty-five days apart if you're more than one year but less than three years out from your first period
- Irregular cycles are less than twenty-one days or greater than thirty-five days apart, or less than eight cycles per year if you're more than three years out from your first period ever and you're not in perimenopause
- Also, if you've had any cycle that's gone beyond ninety days if you're one year out from your first period ever, your cycles are considered irregular

There are some other criteria that are more relevant for your medical team, but that should give you an idea of what is meant by a regular versus irregular cycle.

Points number one and two used to apply to me. My period would sometimes show up after thirty days, sometimes forty-five days, and sometimes it would vanish for months, even when I was well into my twenties. That pretty much sums up the definition and was due to my unmanaged polycystic ovarian syndrome (PCOS), which is the most common cause of erratic and delayed cycles. Now, as mentioned in the Introduction, my period makes a routine appearance every thirty-two days (with some minor deviations, and that's A-OK!).

So, don't fret if your cycles aren't exactly twenty-eight days long. If you're experiencing persistent large deviations from your usual cycle length (especially if it's beyond the twenty-one or thirty-five-day mark), then that's something to discuss with your healthcare team.

CHAPTER HIGHLIGHTS

- The menstrual cycle is the combination of the ovarian cycle and uterine cycle. It consists of four parts:
    - Menstruation, which is your period.
    - Follicular phase, which is when egg-containing follicles and the endometrial lining grow.
    - Ovulation, which is when the dominant follicle ruptures to release an egg.
    - Luteal phase, which is when progesterone is made to mature the endometrial lining followed by a drop in hormones that destabilizes the lining when the body recognizes that pregnancy didn't occur.
- Cycle length is calculated from the first day of full period flow until the day before the start of the next period.
- The average length of a menstrual cycle is twenty-eight days long; however, an established cycle can be anywhere from twenty-one to thirty-five days long.
- Cycle length should be fairly consistent for you, although it can vary now and again based on what's happening in your life. There are times when cycle length can naturally be irregular, like during perimenopause, in the first year after your periods initially start, and while breast milk is being produced.

CHAPTER 3
Phases of the Menstrual Cycle

Before we dissect the phases of the menstrual cycle, it's important to keep in mind that menstrual experiences can vary from person to person. These differences result from a multitude of interconnected factors like genetics, anatomy, age, pre-existing or undiagnosed health conditions, medication use, nutrition status, exercise and movement, sleep, stress levels, trauma, access to clean water and period products, environmental exposures, menstrual education, sociocultural views on periods, and more. This may seem like a lot to think about (and rightfully so!). We'll return to some of these elements in later chapters to help connect some dots and offer practical tips for you to utilize in your life. For now, let's focus on the general sequence of events in the four parts of the cycle and how that can be reflected in your menstrual experience.

Menstruation

Your period marks day one of your menstrual cycle. Flow is instigated by a drop in hormones in the preceding cycle as the body realizes that pregnancy did not occur. This hormonal decline, mainly in the hormone called progesterone, causes your endometrial lining (the innermost lining of the uterus) that had grown to destabilize and shed. Even though menstruation is the culmination of the previous cycle, it's labelled as Cycle Day One, presumably because of how visible a marker your period flow provides.

On average, thirty to fifty millilitres of blood is lost over three to seven days, although some people might naturally sit slightly beyond that volume range.

Menstrual flow contains blood, endometrial tissue, cells from the vaginal wall, secretions from the vagina and cervix, and some of the bacteria that inhabit the reproductive tract. Rarely, a big membranous tissue called a decidual cast can be shed during menstruation. This happens when the endometrial lining is sloughed off intact instead of in small bits. There isn't much data on decidual casts, but case reports point to an association with progestogen-containing oral contraceptive use or discontinuation, as well as ectopic pregnancy.

BOX 2. How much is too much?

It can be a challenge to quantify your blood loss. Many period products will list the approximate amount of blood they hold, which can be a useful way to gauge the amount of blood you lose during your period; however, it can be quite difficult to do so if you change your period products before they're fully saturated or if you're leaking through them.

To give you an idea, five to six millilitres is the approximate amount of blood fully absorbed by one regular-sized pad or tampon. This is about what a teaspoon can hold. You can use this as a general rule of thumb to approximate your blood loss. If you're using a cup or disc, most of them will have volume markers to help you out. Chapter 25 has more information on period products.

Menstrual blood loss tends to be higher on the first one to three days of the cycle, tapering off to smaller quantities as the bulk of the endometrial lining has been shed. As a period ends, the flow tends to decrease gradually over a few days, but it might simply stop abruptly. Some menstruators might even experience on-and-off spotting or light flow in the last couple of days.

Normal period blood loss can look like replacing absorbent period products like pads or tampons at three or more hour intervals. Cups and discs tend to hold larger quantities and are usually emptied less frequently. There are pictorial assessments and other clinical tools that your doctors can use to help gauge blood loss with you as well.

The amount of blood you lose as an individual should be fairly consistent from cycle to cycle as we expect your hormone production to lead to a similar amount of endometrial growth (and shedding). There are times when bleeding patterns can be variable naturally, like during puberty, perimenopause, and postpartum. There might be some temporary changes at times based on what's happening in life, like if you recently started a medication, had a bad infection, or were going through something stressful, etc. Then, there are changes that we want to address more robustly.

Heavy or prolonged menstrual bleeding, which is categorized under the term abnormal uterine bleeding (covered in Chapter 18), is when blood loss goes beyond eighty millilitres or seven days of bleeding. This should be discussed with your healthcare team to elucidate the cause, which could be something like fibroids, adenomyosis, a clotting disorder, or endometriosis. Signs that you may have heavy menstrual bleeding can include anaemia, menstrual blood clots that are larger than 2.5 centimetres (one inch) in diameter, frequent period product change, doubling up on period products, and flooding beyond your period products.

Scant bleeding is defined as less than ten millilitres of blood loss spanning all of your bleeding days. Scant bleeding is commonly seen with low hormone production and anovulatory disorders where ovulation isn't occurring. It is also often seen in hormonal intrauterine contraceptive users.

Menstrual blood will be a shade of red—anywhere from a pinkish red to a dark burgundy. There may even be brownish spotting, especially at the tail end of the period, as the uterus is expelling the last bits of lining and the blood is getting oxidized.

Your flow will likely be fairly fluid: not thin like water, but not thick like honey. You may notice some mucus and even some translucent tissue bits. Some small or medium clots may be present, and those are typically not concerning. Clot size is approximated as follows:

- small clots are dime-sized or smaller (a dime has a diameter of 1.8 centimetres)

- large-sized clots are loonie-sized or bigger (a loonie is a Canadian one-dollar coin with a 2.6-centimetre diameter)
- medium-sized clots are somewhere in between (like quarter-sized)

Consistently expelling large clots, especially with other symptoms like heavy bleeding or pain, should be discussed with your doctor. Significant clotting could indicate an underlying condition. A large clot, especially if accompanied by significant pain and a delay in your next period, could even be a spontaneous pregnancy loss. An hCG test, i.e., a pregnancy test, at the time of the clot passing, may be able to shed light on whether that is the case or not, but only if done within a certain time frame.

There's a huge misconception that periods are supposed to be quite painful. This is simply not true.

The endometrium has two layers, the basal layer and the functional layer. It's this latter functional one that grows and is subsequently sloughed off each cycle. This layer has immune cells and blood vessels. When it's being shed, an inflammatory response takes place. This inflammatory process is similar–but different–to what you experience when you injure yourself: there's redness and swelling from increased blood flow, as well as compounds released by the immune system to repair tissue and eliminate waste. During your period, there's more blood flow, more fluid, and numerous immune compounds called prostaglandins and leukotrienes being released for repetitive tissue injury and repair to occur in the endometrium. All of this is an orchestrated physiological process and happens without loss of function or scarring left behind ... It's pretty cool if you ask me!

It also explains why it can be quite normal to experience fluid retention, feelings of pressure or heaviness, and even a bit of discomfort in the pelvis, vulva, or lower back when you're menstruating.

In fact, the onset of period flow can bring along with it many changes. The natural drop in hormones that instigated your period can lead to a bit of low energy starting before the onset of flow that continues into the first couple of days of your period. Fatigue might be worse for those of you with anaemia or heavy bleeding. Your appetite and digestion may change mildly, both of which can impact how you feel.

It's quite usual for menstruators to change up some of their habits and routines during their period to accommodate the changes at its onset. I know that I like to take it a bit easier on day one of my cycle if possible as I tend to feel more tired that day. If you have severe premenstrual syndrome (PMS) or premenstrual dysphoric disorder (PMDD), menstruation likely brings with it significant relief from symptoms that were disrupting life in the luteal phase. On the other hand, distress might be your experience if you have unsettling symptoms like severe pain, nausea, vomiting, headaches, and heavy flow.

Significant pain, taking heroic doses of painkillers, super heavy flow, vomiting, fainting, and missing out on work, school, or life, warrants a proper assessment and support even in the face of the normalization of these symptoms. Chapter 15 is dedicated to painful periods, including causes and treatment, while Chapters 18 and 20 include causes of painful periods and abnormal uterine bleeding like fibroids and endometriosis.

Another misconception is that you're not "supposed" to have penetrative intercourse during a period. There is no medical basis for this; however, there might be cultural or religious taboos as well as personal preferences surrounding the subject. This is a good place to remind you that it is, indeed, possible to get pregnant while you're on your period, although it's unlikely. It all depends on how close together (the end of) your period and ovulation are! This will make more sense when ovulation is covered later in the chapter.

Follicular Phase

The follicular phase technically starts with the period; however, it's easier to keep its hormonal shifts separate from the process that is menstruation.

This is when follicle-stimulating hormone (FSH), well, stimulates the growth of follicles. People assigned female at birth enter the world with millions of follicles, which are sacs that contain immature eggs in them. A small group of these are recruited every cycle. One follicle becomes the dominant follicle, and this is the sac that will rupture at ovulation time, marking the end of the follicular phase. The other non-dominant follicles that had been recruited for that cycle undergo a kind of disintegration process called atresia.

Follicle-stimulating hormone is produced in response to a brain hormone called gonadotropin-releasing hormone (GnRH). GnRH is a fascinating hormone. Scientific research in animal models, corroborated in human case studies, has shown that it is released in fast and slow frequency pulses. During the follicular phase, GnRH is secreted in slower frequency pulses, which causes follicle-stimulating hormone production. As follicles grow and we move deeper into the follicular phase, GnRH pulses start to speed up, and these faster releases will cause luteinizing hormone to be produced (more on that later).

As follicles grow under the influence of FSH, estrogen also rises. Estrogen, in turn, supports FSH in its roles and stimulates the endometrial lining to grow—it's your builder hormone. This way, there's a place for the fertilized egg to be implanted if fertilization occurs. One of estrogen's other jobs is to instigate cervical mucus production.

The cervix has tons of glands that produce mucus consisting of proteins, electrolytes, sugars, fatty acids, and enzymes. Cervical mucus production is a normal part of the menstrual cycle. It indicates that you have adequate estrogen and an appropriate response to estrogen leading up to ovulation. There's no single amount that's the "correct" amount to be making; different people make different quantities. The quantity or change in quantity over time can give us some clues, though, especially if there are other changes like those in libido, mood, energy, and the vulvovaginal areas. Some medications and conditions can affect your hormones or overall mucus production, like low hormone conditions and antihistamines used for allergies that decrease mucus. Then, there are also low-estrogen life transitions that may have manifestations like when one is lactating or in the menopause transition.

Cervical mucus can be white, off-white, or clear, and its characteristics can change throughout the cycle under the influence of fluctuating hormones. The most typical progression in the follicular phase is to have none to little mucus after a period, especially if you have a longer follicular phase. As ovulation nears and estrogen levels rise more steeply, there is an accompanying increase in mucus production. It becomes more slippery, stretchy, and alkaline to facilitate sperm travel and survival. This is why it's called fertile cervical mucus.

It is totally normal for cervical mucus to have a slight odour, especially as it transits through the vaginal tract. It mingles with sweat, sebum, and fluid secreted in the vagina and vulva. It's not a bad smell ... it's simply a distinct smell. If you're noticing an odour that's fishy or foul, especially if accompanied by yellow, green, or crumbly discharge as well as symptoms like itching, burning, or pain, that could indicate an infection. This definitely constitutes a reason to chat with your doctor and get appropriate treatment. Box 1 in Chapter 1 covers vaginal odour in more detail.

The follicular phase ends with a sustained rise in estrogen and high-frequency pulses of GnRH that lead to a big spike in luteinizing hormone and ovulation.

Ovulation

Even though the period gets more attention, ovulation is the real star of the show that is your menstrual cycle. It takes a coordinated effort to release an egg every cycle. The occurrence and timing of ovulation are critical to cycle length, bleeding patterns, fertility (there's no chance of pregnancy without an egg!), progesterone production, luteal phase experiences, uterine cancer risk, and more.

Ovulation occurs when luteinizing hormone (LH) rises sharply at the end of the follicular phase. This surge is a result of high-frequency GnRH production in the brain as well as estrogen levels reaching a certain threshold for two days. The cool thing about these hormonal increases is that rising GnRH, estrogen, and LH all happen in an interconnected fashion! They all influence one another.

As luteinizing hormone levels surge, the dominant follicle reacts by releasing its egg. If you've ever used or heard of an ovulation predictor kit, LH is the hormone it's detecting, so you'll get the highest or clearest reading just before ovulation.

Once the egg is out of its sac, it will travel through the fallopian tube into the uterus. It may even meet up with sperm along this route if there are any around—this will lead to fertilization.

Apart from cervical mucus production, the cervix also changes around ovulation time: it becomes a bit softer, more open, and retreats higher (closer to the body of the uterus). This can be used as a sign of fertility along with other parameters covered in Box 3 whether you're trying to conceive or avoid pregnancy ... or are simply curious! Ovulation signs, especially cervical mucus, can help pinpoint your fertile window. This is the timeframe during which conception can happen. The fertile window begins approximately five days before ovulation, as sperm can survive for three to five days. The fertile window ends approximately one to two days after ovulation, based on the lifespan of an egg.

Now, let's return to the question of whether pregnancy can occur with unprotected intercourse during a period. It all depends on how long your flow lasts and when you ovulate. For example, if you have unprotected intercourse on cycle day six and then ovulate on day nine, fertilization could occur—it really has to do with the viability of sperm in the body (which is three to five days), the timing of ovulation, and when intercourse happened. Menstrual blood actually creates an environment that is conducive to sperm travelling and surviving fairly well! So, contraception should be a consideration if pregnancy prevention is a goal of yours. The oral contraceptive pill works by suppressing hormones and ovulation so that there's no egg to be fertilized to begin with. Contraception, including fertility awareness-based methods, is discussed in Part Three.

The empty shell left behind from the follicle is called the corpus luteum. This is the structure that produces your other major sex hormone called progesterone. It also makes estrogen. We'll return to this structure in the luteal phase section later.

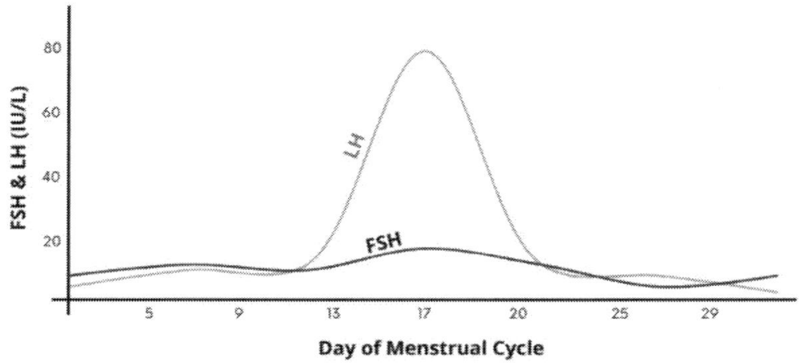

DIAGRAM 5: *Graphical representation of LH and FSH throughout a 32-day menstrual cycle. LH levels are volatile, with a huge spike in the middle of the menstrual cycle to trigger ovulation. FSH levels vary more during the cycle, increasing during the follicular phase for follicle growth and maturation, decreasing slightly when LH begins to rise, increasing mid-cycle for follicle preparation, decreasing again in the early luteal phase, and increasing one more time at the end of the luteal phase to recruit follicles for the next cycle.*

When luteinizing hormone surges, estrogen dips a bit, and these hormonal changes can show up in various ways. Some menstruators don't notice much during this time, but some of you may experience fatigue, bloating, headaches, and digestive changes. Some of you may even experience *mittelschmerz,* which is a German word that translates to *middle pain,* referring to transient pain typically felt in the lower abdomen as ovulation is occurring. This can look like an ache, twinge, or spasm. Some symptoms, especially if severe, new, or worsening, could even indicate an underlying issue. For example, mid-cycle spotting and severe or lingering mid-cycle pain may be indicative of endometriosis or adenomyosis.

Ovulation can happen on different days for different people. It does not necessarily happen on day fourteen or in the exact middle of a cycle. Most period apps guesstimate ovulation based on an algorithm that pools data from users who use that app (unless, of course, you're using a more advanced fertility or cycle monitor that uses objective and personalized data). Ovulation happens when the hormonal milieu of the late follicular phase instigates the rupture of the dominant follicle, which might even vary slightly from cycle to cycle. I usually ovulate some time between cycle days seventeen and nineteen, which is in line with my cycle being approximately thirty-two days long. Tracking and charting your cycle by paying attention to the signs of hormone change can give you valuable information about when you, as an individual, ovulate.

A cycle where ovulation does not occur is called an anovulatory cycle. This may show up as irregular bleeding patterns and/or inconsistent cycle lengths. Ovulation is what determines the length of a cycle.

Anovulatory cycles can be expected around puberty when menstrual cycles are being initiated and the body is learning how to mount a regular menstrual cycle. They're also common during times of lactation (i.e., milk production), and in the menopause transition as the final menstrual period ever approaches. There are also other causes of anovulatory cycles, like my personal diagnosis called polycystic ovarian syndrome (PCOS, discussed in detail in Chapter 19); PCOS is the most common condition leading to anovulatory cycles.

Having ovulatory cycles with fluctuating hormones during your reproductive years has been shown to be good for your health not only now but also in the future. Having an ovulatory menstrual cycle tells us you have the capacity to mount a coordinated hormone pattern, whereas a lack thereof may indicate a lack of resources or something else going on. In fact, high stress, inadequate nutrition, eating disorders, and hormonal disturbances can all throw off this cyclical rhythm. Cyclical fluctuations of estrogen and progesterone have an impact on cardiovascular and bone health as well. There are conditions, however, where these hormone fluctuations can be problematic; a great example is premenstrual dysphoric disorder, in which luteal phase estrogen and progesterone lead to severe mental-emotional distress.

BOX 3. How do I tell if I'm ovulating?

The body can give off various signs and signals that ovulation is occurring. Sometimes, an attempt to ovulate can look like ovulation even though no egg is released (I know, it can be very confusing!).

Tracking aspects of your cycle and employing additional testing through your healthcare provider can be a helpful way to determine your fertile window, a practice traditionally reserved for (in)fertility cases, but accessible to most of us to understand our cycles and utilize this information for contraception as well. Tracking at least three cycles, but ideally, six to twelve is important as this is the only way to establish patterns, something that can't be done with data from one cycle alone. Cycle tracking is discussed in more detail in Chapter 7.

Here are the signs that might indicate you're ovulating:

1. Cervical Mucus: Stretchy, slippery, egg-white-like mucus that is present in high quantities starts a few days before ovulation and can indicate that the fertile window has started. After ovulation, when progesterone rises, this cervical mucus usually becomes thicker and tackier, marking the end of the fertile window. This is the sign that fertility practitioners rely on the most when charting menstrual cycles. Chapter 11 has more details.

2. Basal Body Temperature (BBT): After ovulation, body temperature rises as a result of progesterone's thermogenic, i.e., heat-producing, effect. It may increase your temperature by 0.5 to 1 degree Fahrenheit or 0.3 to 0.5 degrees Celsius. Taking the body's temperature orally, vaginally, or rectally upon waking before getting out of bed or consuming anything at approximately the same time daily is how to start mapping out your BBT. This offers a retrospective look at ovulation since temperature rises *after* ovulation. Please note that stress, sleep disturbance, room temperature, waking time change, recent start or discontinuation of the birth control pill, fever, and alcohol can all alter BBT, so it's not entirely reliable on its own but may be combined with other signs to give you an idea of ovulation.

3. Ovulation Predictor Kits (OPK) or LH Strips: These look kind of like pregnancy tests and measure your luteinizing hormone (LH), which spikes right before ovulation. It's a urine test that will give you an indication of elevated LH levels through a line, symbol, or number. The surge in luteinizing hormone can precede ovulation by thirty-five to forty-four hours, with levels peaking ten to twelve hours before

ovulation. Note: some conditions, like PCOS, can lead to baseline elevations in LH, which can render these tests unreliable.

4. Cervical Position: This technique is a bit more involved and might not be feasible for some people. It requires the insertion of fingers into the vaginal tract to feel what the cervix is like before, during, and after ovulation. The cervix becomes softer and shorter under the influence of mid-cycle hormonal changes.

5. Cycle patterns: If your cycles are fairly regular, you have within-range blood loss, and you notice fertile cervical mucus around ovulation time, these signs likely, not definitely, indicate ovulatory cycles. ovulation signs like fertile cervical mucus, then you have some more evidence. The corpus luteum that makes progesterone only lasts for ten to twelve days, so we expect a period approximately two weeks after an egg is released.

6. Other Signs: Changing hormones at ovulation time can lead to manifestations such as fluid retention, digestive change, and ovarian pain or mittelschmerz. Pay attention to these as they can indicate ovulation in combination with the other methods listed here.

7. With assistance from a doctor, there are other ways to confirm ovulation. These methods are typically employed in fertility clinics and include post-ovulatory progesterone testing as well as imaging. There are at-home progesterone tests that have entered the market recently as well.

Luteal Phase

Ovulation is the dividing line between the follicular and luteal phases. The luteal phase is sometimes divided further into the early and late luteal phases. The early luteal phase is defined by an increase in both estrogen and progesterone, while the late luteal phase is when hormones are on a downward journey as depicted in Diagram 6.

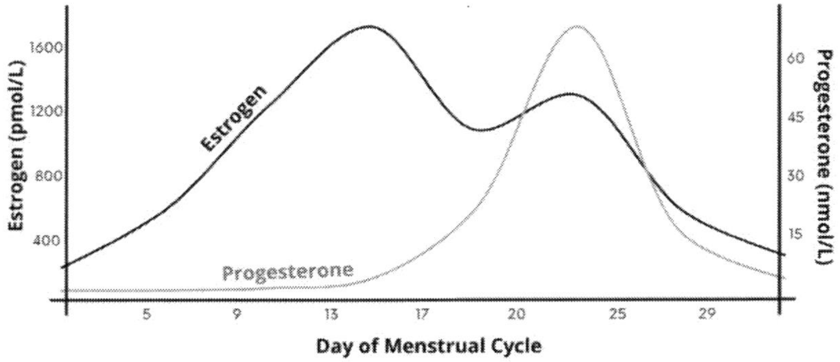

DIAGRAM 6: *Graphical representation of estrogen and progesterone throughout a 32-day menstrual cycle. Please note the different units of estrogen (pmol/L) and progesterone (nmol/L), which is a 1000-fold difference. Estrogen rises at the beginning of the cycle, dipping shortly before ovulation. It increases after ovulation until the body registers that pregnancy did not occur. Progesterone levels are low until after ovulation, at which time it is produced by the corpus luteum. Its sharp decline at the end of the luteal phase triggers the endometrium to shed during menstruation.*

Progesterone is produced by the follicle shell called the corpus luteum which is left over after the egg is released. It peaks six to eight days after ovulation, with luteinizing hormone influencing its production. Progesterone can lead to a few obvious changes like a mild increase in basal body temperature (BBT) and cervical mucus that is thicker, tackier, and more acidic. This may act as a kind of barrier protection in case the egg was fertilized–as a natural contraceptive since sperm can't live and move as well in that environment compared to more fertile cervical mucus.

A process called decidualization of the endometrium begins in the luteal phase, whereby the endometrial lining matures and transforms into a more receptive medium for a potential pregnancy. Both estrogen and progesterone are needed for this activity, but progesterone does the heavy lifting. It stabilizes the endometrial lining, recruits immune cells, and stimulates blood vessel growth. The decidualization process is crucial for pregnancy and placental development. If a pregnancy occurs, then decidualization continues and the corpus luteum sticks around to secrete progesterone and estrogen as well as a protein called inhibin A to keep the endometrial lining thick, prevent contractions of the uterus, and block the recruitment of follicles for the next cycle.

In the absence of conception, however, decidualization is halted as the corpus luteum starts to deteriorate. The corpus luteum can only last for ten to twelve days, so once ovulation occurs, a period will start in approximately two weeks. This can offer a confirmatory retrospective on when ovulation happened. There are times when this kind of retrospective is inaccurate, like when people have something called a luteal phase defect. This is when the corpus luteum is not making adequate progesterone or the uterus isn't responding to progesterone, and can manifest as fertility issues, premenstrual spotting, recurrent pregnancy loss, short menstrual cycles, and more.

Under regular circumstances, the corpus luteum deteriorates in the late luteal phase and is now called the corpus albicans. Progesterone and estrogen, which were being produced by the corpus luteum, begin their descent. All of this triggers a cascade of events that involves the recruitment of follicles for the next cycle thanks to follicle-stimulating hormone, a destabilization of the uterine lining, immunological changes, inflammation, and the initiation of menstrual bleeding to shed the lining that had built up so that growth can begin anew.

There may be other manifestations of hormonal shifts in the luteal phase, especially in the seven(ish) days leading up to a period: fluid retention, cravings, mood changes, digestive differences, and fatigue are common ones. Mild changes are usually a product of changing hormones and are deemed normal. For example, your body might need anywhere between thirty(ish) to over two hundred calories more in the luteal phase depending on your resting metabolic rate and activity levels. This can show up as increased appetite and cravings, or even poorer exercise tolerance as fuel demand has increased even though you're eating the same as you normally do. Your fluid dynamics and electrolyte status might be impacted, especially if you're active, so paying special attention to hydration can be an important part of your strategy here.

Then, there might be not-so-mild changes like volatile mood swings, migraines, and/or feeling like you can't cope with life. If you're struggling with physical, mental-emotional, and/or somatic symptoms for one to two weeks per month, I encourage you to seek out help. Chapters 16 and 17 on PMS and PMDD lay out some insight and options for you.

Hormonal shifts throughout the menstrual cycle are experienced differently from person to person. Not only that, our brains can perceive hormonal shifts differently before ovulation and after ovulation (which is fascinating but can be frustrating!). Then, there are changes that slowly occur over time as we age and go through life.

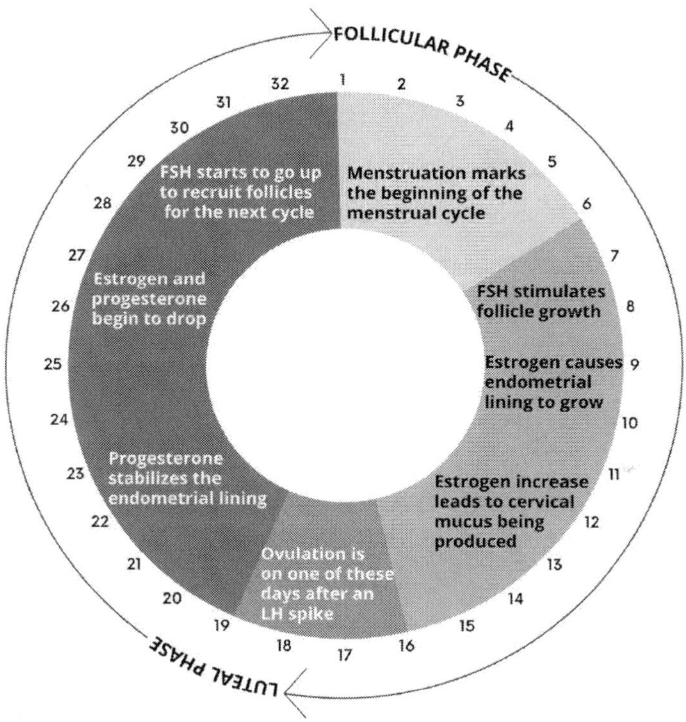

DIAGRAM 7: *Cyclical representation of key events in a 32-day menstrual cycle. Day 1 of the cycle is when menstruation begins. A typical period lasts between three and seven days. Menstruation also marks the beginning of the follicular phase, during which FSH causes follicular to grow and a dominant follicle matures. At the same time, estrogen causes the endometrial lining to grow. LH spikes approximately halfway through the cycle and causes ovulation to occur, marking the beginning of the luteal phase. Progesterone is then produced by the corpus luteum to stabilize and mature the endometrium. After a few days, when the body recognizes the lack of pregnancy, hormones begin to drop, which destabilizes the endometrial lining. FSH starts to go up to recruit follicles for the next cycle as estrogen and progesterone decline.*

The fact is: that our lives, environments, and hormones are intricately connected and dynamic parts of a whole. We are coming to understand that hormones happen to us, and we happen to our hormones simultaneously. There is extensive interplay between childhood experiences, lifestyle, stress, genetics, life experience, nutrition status, access to resources, geography, social supports, and other factors when it comes to how symptomatic or asymptomatic menstrual cycles are, especially in the late luteal phase. Even though the premenstrual time can come with its challenges, there are ways to smoothen the ups and downs (more on that later!).

Back to the Beginning

Once the luteal phase ends, a new menstrual cycle begins with a shedding of the uterine lining as a menstrual bleed.

One phase leads to the next. There is no luteal phase without the follicular phase and vice versa. It is a sophisticated hormonal dance that, when executed properly, leads to this very important cycle of ours; a dance with estrogen and progesterone as its main stars along with many other major and minor parts.

CHAPTER HIGHLIGHTS

- Menstruation marks the beginning of a menstrual cycle, with the average blood loss amounting to thirty to fifty millilitres over the whole period, which can take three to seven days.
- There are inflammatory compounds involved in the process of menstruation which can lead to feelings of pressure, heaviness, and discomfort. That being said, periods are not supposed to be painful, especially to the point where they are interfering with your life.
- The follicular phase (and a new menstrual cycle) technically begins with the period but is typically discussed separately for the sake of ease. In the follicular phase, follicle-stimulating hormone (FSH) is released, egg-containing follicles grow, a dominant follicle is selected, estrogen rises, cervical mucus is produced, and the endometrial lining builds up.

- Ovulation occurs when the dominant follicle releases an egg thanks to a surge in luteinizing hormone (LH). Some ways to tell if ovulation has happened include:
    - Mittelschmerz or mid-cycle pain
    - High amounts of egg-whitey cervical mucus
    - Increase in basal body temperature (BBT) after ovulation
    - Positive ovulation predictor kit test that measures LH
    - Retrospective confirmation based on a period starting approximately two weeks after ovulation
    - Imaging or progesterone testing with a healthcare provider or fertility clinic
- Cervical mucus is clear or whitish and changes throughout a menstrual cycle. Different people make different amounts. Cervical mucus produced around ovulation time is alkaline and less viscous to facilitate better sperm travel and survival. After ovulation, progesterone causes the mucus to be thicker and more acidic.
- The early luteal phase is marked by high levels of estrogen and progesterone made by the corpus luteum and leading to the stabilization and maturation of the endometrial lining. In the late luteal phase, when the body realizes that pregnancy did not occur, the corpus luteum starts to disintegrate, causing a drop in progesterone levels. This leads to the destabilization of the endometrial lining and a new menstrual cycle.
- Everyone has their own menstrual experience that is impacted by a variety of different factors such as genetics, age, medical history, access to resources, environment, and family of origin. It's normal for cycles to change over time.

CHAPTER 4
Main Menstrual Hormones

This chapter is all about the biochemical brigade that is your menstrual hormones. You might be thinking that this is seeming more and more like a textbook with a full chapter dedicated to hormone science, but this is the only way to appreciate the sophisticated communication system they are all part of and how the different pieces fit together. You'll also get a taste of the sheer breadth of what your hormones do, how some common menstrual conditions develop, when to time certain hormone tests, and why we care about these hormones in the first place.

Let's begin with what hormones actually are. They are chemical messengers that tell your cells to do or not do something. Most of them attach to binding sites on the surface of your cells or inside your cells to initiate the responses that are under their purview. For example: estrogen binds to estrogen receptors in the endometrium and causes the endometrial lining to grow. (It does a lot of other things as you'll shortly learn!)

The menstrual cycle is governed by hormones, with communication happening between the ovaries and two parts of the brain called the hypothalamus and pituitary gland. The trinity of these organs is called the hypothalamic-pituitary-ovarian (HPO) axis. The coordination among these organs is critical to a menstrual cycle ... and, like any relationship, miscommunication can lead to many problems!

There are five main hormones involved in the HPO axis, and those are the ones we'll discuss first:

- Gonadotropin-releasing hormone (GnRH)
- Follicle-stimulating hormone (FSH)
- Luteinizing hormone (LH)
- Estrogen
- Progesterone

Gonadotropin-releasing hormone (GnRH)

GnRH is the conductor of the symphony that is your menstrual cycle. It's produced and released by the hypothalamus in the brain. It is an aptly named hormone because it instructs the pituitary gland to release your gonadotropin hormones, namely FSH and LH.

It's a pretty interesting (and slightly enigmatic) hormone. As discussed previously, it's released in pulses; the frequency of these pulses–i.e., fast or slow–will dictate which hormone is released. You have to admit, the body is pretty neat!

At the end and into the beginning of a menstrual cycle, GnRH pulses are slow, which leads to follicle-stimulating hormone production. Deeper into the follicular phase, the frequency rises and leads to a surge of luteinizing hormone right before ovulation.

One of the hormones that will dictate how fast GnRH is released is called *kisspeptin*. We're learning more and more about this hormone over time. We know that it is crucial to establish a menstrual cycle during puberty and for menstrual cyclicity. It also seems to play a role in thyroid function and blood sugar regulation. In fact, it is quite responsive to energy availability in the body. This kisspeptin and another hormone called leptin are both responsive to energy availability in the body. Having your caloric needs met is critical to having a healthy menstrual cycle, and these hormones are involved in that; if there's a lack of fuel, the brain will divert resources toward survival instead of reproduction. This concept is covered under hypothalamic amenorrhea in Chapter 21 and is something we might see as part of relative energy deficiency in sport (RED-S).

FSH (Follicle-Stimulating Hormone)

FSH is released from the anterior pituitary gland in the brain as a result of low-frequency pulses of GnRH.

Once the body realizes you're not pregnant, FSH starts to rise at the end of the luteal phase before menstruation even starts to recruit follicles for the upcoming cycle.

In the follicular phase, FSH leads to follicle growth and maturation. Other jobs under its purview include estrogen production by activating the synthesis of an enzyme called aromatase. This enzyme will convert androgens like testosterone and androstenedione to estrogen.

FSH also increases receptors for the ovulatory hormone, LH; this way, LH can bind to these receptors to do its thing. As a dominant follicle grows each cycle, the hormones it produces will signal to the brain that it's time to decrease FSH and make room for LH to shine.

Too little FSH can indicate an issue with the hypothalamus or pituitary glands. Too much can mean that your organs aren't responding appropriately or a growth/tumour might be causing overproduction. In the final few years of the menopause transition and post-menopause, FSH can naturally be elevated as menstrual cycles are coming to an end or have ended.

LH (Luteinizing Hormone)

Like follicle-stimulating hormone, LH is also made by the pituitary gland, but it's produced as a result of high-frequency pulses of GnRH at the end of the follicular phase. If you've ever used a urine ovulation strip called an ovulation predictor kit, LH is the hormone being measured.

Apart from GnRH, estrogen plays a role in LH secretion as well. Low levels of estrogen will stop LH from being produced in a significant amount. Once the ovaries have produced a critical amount of estrogen for approximately two days, LH can soar high and the egg can fly out of the follicle and into the fallopian tube. Actually, another one of LH's jobs is to weaken the ovarian wall so that the egg can pass through the ovary and into the fallopian tube for potential fertilization. Its other major role is to stimulate estrogen and progesterone production from the corpus luteum.

Low LH levels, similar to low FSH levels may be due to issues with the brain-ovary connection. High levels, especially when sustained as seen in PCOS, can interfere with follicle growth and ovulation in general.

Diagram 8 is a simplified version of the HPO axis showing the hormonal communication between the brain and ovaries that results in a menstrual cycle.

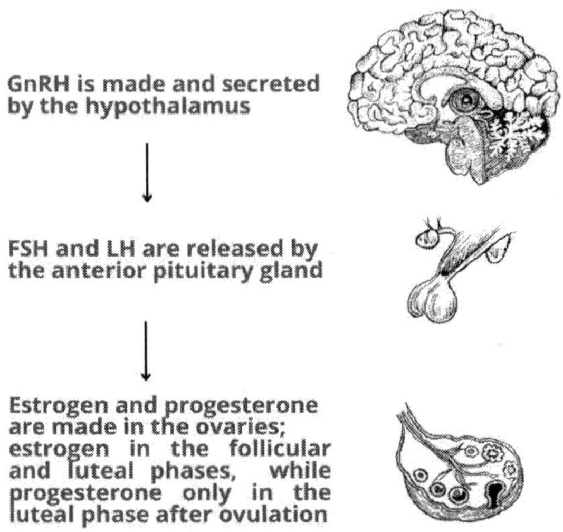

DIAGRAM 8: *Hypothalamic-pituitary-ovarian (HPO) axis). GnRH released by the hypothalamus in the brain relays a signal to the pituitary gland (also in the brain, shown enlarged separately for convenient visualization). The pituitary gland releases FSH (slow-frequency GnRH pulses) and LH (high-frequency GnRH pulses). In the ovaries, FSH will cause follicle growth and maturation and estrogen to go up. Once LH is released, ovulation will occur in the ovary and progesterone will be released by the corpus luteum.*

Estrogen

Estrogen is a hormone that many people love to hate and blame for what it does even though it is a critical hormone for not only your menstrual health but also your metabolism, bones, brain, heart, skin, and so much more.

There are four different kinds of estrogen: estrone (E1), estradiol (E2), estriol (E3), and estetrol (E4). Estradiol is the most potent and abundant estrogen during our reproductive years, so that's the one we'll mainly be discussing in this book, and I'll be using the general term estrogen in place of estradiol.

The majority of your estrogen is made from testosterone using an enzyme called aromatase. When you break estrogen down for elimination, most of it happens in the liver and digestive tract.

As estrogen levels slowly start to increase at the beginning of the menstrual cycle, luteinizing hormone (LH) production is inhibited so that follicle-stimulating hormone can grow follicles; estrogen helps with that maturation process. It also stimulates endometrial cells to multiply and progesterone receptors to be expressed for when progesterone needs to do its thing.

Estrogen levels continue to rise into the later part of the follicular phase, reaching a critical threshold for approximately forty-eight hours. This contributes to a faster release of GnRH and a subsequent surge in LH, triggering. Estrogen takes a bit of a dip at this time and then has a resurgence after the egg is released.

Estrogen has other reproductive roles in the body: it's responsible for breast and mammary duct development during puberty and pregnancy, it supports the proliferation of vaginal and vulvar cells, and it increases the expression of progesterone receptors for the luteal phase.

As estrogen starts to naturally decrease in the years leading up to menopause, shifts in vaginal health can be seen, including a thinning of the vaginal walls, dryness, irritation, and low lubrication. These symptoms can also be experienced in other low-estrogen states like during lactation or if you're on anti-estrogen meds. Lubricants and topical estrogens can be used for vulvovaginal changes like those seen in genitourinary symptoms of menopause (GSM).

As mentioned earlier, estrogen has an impact beyond the reproductive system. It's important for healthy cholesterol, fat metabolism, and blood sugar regulation. Its availability throughout your life contributes to a lower risk of atherosclerotic heart disease (the kind where plaque builds up in the blood vessels). It's involved in how and where fat is deposited (usually around the hips in reproductive years) and it has an impact on our metabolic rate.

It helps with cortisol release, regulating how much cortisol is able to do in the body. And, it promotes bone health, mainly by preventing bone loss, which becomes especially relevant with the natural decline in estrogen as menopause is reached.

Times when we would consider giving someone estrogen as a therapy (typically with progesterone) include during perimenopause for hot flashes and night sweats, early or surgical menopause to prevent the negative cardiovascular and bone outcomes related to an early loss of sex hormones, male-to-female trans support, fertility purposes, and as a form of contraception (think of the pill and patch).

Estrogen has gotten a bad rap over the years, and it has been blamed for a plethora of reproductive issues. Some of this stems from a good ole case of oversimplification in how information is presented in the online space. Some of it is due to the prevalence of estrogen-sensitive breast cancers. The ups and downs of perimenopausal estrogen leading to unpleasant symptoms as well as the overall decline in estrogen around that time contribute a bit. Some from the use of estrogens found in the birth control pill and its side-effect profile. Some of it is due to the historic use of estrogen therapy without pairing it with its partner, progesterone. Then, there are conditions like endometriosis that are impacted–not caused–by estrogen.

Estrogen also has a complex relationship with histamine–the immune compound associated with allergies–that can contribute to inflammatory symptoms like cyclic migraines, general malaise, fatigue and hormone-based conditions, particularly in the luteal phase.

There are also xenoestrogens–foreign estrogen-like compounds–we are exposed to regularly in life that have very real physiological effects on our hormonal and metabolic health. Endocrine disruptors, many of which look kind of like estrogen, can negatively affect us even when the exposures are small (more on that in Chapter 30). We've only scratched the surface when it comes to how chronic exposure to hormone disruptors is affecting our health and that of future generations. Individual differences in how estrogen impacts the body with respect to production, utilization, and elimination probably account for some of the confusion as well, especially when we cannot pinpoint those elements definitively given the tools we have right now.

Add misinformation and extrapolations to the mix, and you have a recipe for mixed feelings about this crucial hormone.

The thing is, we *need* estrogen for, well, health and life. I hope that this section helps you appreciate estrogen for all that it does! Yes, we need to metabolize and eliminate it properly while decreasing our exposure to hormone disruptors, but we *like* and *want* estrogen to be around, especially with its partner in crime, progesterone.

Progesterone

Progesterone is made in the ovaries, corpus luteum, adrenal glands, and placenta. The primary way to regularly make progesterone is through ovulation. When we don't ovulate, we don't make much progesterone.

Progesterone, like estrogen, is an important hormone for your menstrual and overall health. It has numerous roles in the body: it stabilizes the endometrial lining in the luteal phase, facilitates angiogenesis (new blood vessel formation), decreases uterine contractions, promotes blood flow to the uterus, increases the process of bone formation, crosses the blood-brain barrier to support brain cell survival, and has anti-inflammatory properties (it is when progesterone drops before a period that a local inflammatory response is initiated). It also affects electrolyte balance, gut motility, insulin release, and metabolic rate.

It's the hormone primarily responsible for the decidualization of the endometrium, i.e., the structural changes associated with the maturation and stabilization of the endometrial lining in preparation for implantation and pregnancy. During pregnancy, progesterone levels are high to disallow the uterus from contracting, inhibit lactation, promote blood flow to the uterus, and allow the immune system to be tolerant to an embryo, among other roles.

Progesterone causes a temperature rise (increases set point by around 0.3-0.7 degrees Celsius) if you make it in adequate amounts, which is why basal body temperature can rise after ovulation. It also makes your cervical mucus thicker, marking the end of your fertile window.

Progesterone works in concert with *and* in opposition to estrogen; like any solid partnership, we need them both! If you don't have enough estrogen or progesterone during adolescence, your peak bone mass (the maximum amount of bone you can have) is decreased. A lack of progesterone can impact cell growth, leading to excessive growth of the uterine lining called endometrial hyperplasia as well as breast cell growth as it's not there adequately to oppose the action of estrogen. Uncontrolled cell growth puts us at risk of cancer. This is a major reason why progesterone is given for menopause hormone therapy with estrogen for those who still have a uterus–to prevent excessive growth of the uterine lining.

Progesterone can be given as a therapy for a variety of reasons, like in perimenopause (alone or with estrogen). It can also help people with recurrent pregnancy loss or unexplained infertility carry to term.

There are also progesterone-like compounds called progestins. These compounds can offer contraception on their own (like the progestin called levonorgestrel in the hormonal IUD, discussed in Chapter 10) or in combination with estrogen (like in the oral contraceptive pill, discussed in Chapter 9).

The term progestins always refers to non-bioidentical progesterone, meaning they are progesterone-like compounds but do not have exactly the same chemical structure. There are many different kinds of progestins which have similar effects to progesterone, like thinning the uterine lining and thickening cervical mucus. However, some progestins can have very different effects on other parts of the body (which is why we may choose a certain kind based on the desired outcome). As a group, progesterone and progestins altogether are called progestogens.

Other Hormones

Your hormones don't work in isolation; in fact, nothing in the body works in isolation. Honestly, we still have much to learn when it comes to the intricate nature of hormonal interplay. Hormones are typically studied in isolation first, then in relation to other hormones, a disease, or question we want to answer, and then we have to hold all of that in context of the rest of the body and our environment. That's how we've come to learn about all of the hormonal players involved in the menstrual cycle.

There are many hormones beyond the primary ones we've just discussed that have important roles. The impacts of these hormones become more apparent when they're out of range. Some of these you've most likely heard about (like testosterone, thyroid hormones, and cortisol), while others are lesser known (like DHEA and insulin).

Testosterone

Similar to estrogen, testosterone is a hormone that gets a bad rap. It is the body's primary androgen. An androgen is basically a hormone that is associated with traits belonging to those assigned male at birth, like terminal (thick, coarse) facial hair and a deeper voice.

Testosterone's roles include healthy bone metabolism, providing a precursor to your most potent estrogen called estradiol, regulating blood calcium levels, building and maintaining lean mass, mood, mental health, sex drive, and more.

You need adequate testosterone to function and be well. Low testosterone can manifest as fatigue, poor muscle recovery, low libido, and low mood. If you have too much circulating testosterone, which can be seen in conditions like PCOS, it can interfere with menstrual cycles, impairing ovulation and the hormonal feedback loop needed for healthy and regular cycles. More on this in Chapter 19. Elevated levels may also lead to symptoms like acne, hair loss, and facial hair growth. The birth control pill is often used for these symptoms as it not only leads to regular bleeds but also decreases circulating free testosterone which leads to hair and skin issues.

DHEA

DHEA stands for dehydroepiandrosterone. It's usually measured in blood as DHEA sulphate (DHEA-S). It's your other primary androgen and is made largely by the adrenal glands, organs that sit atop the kidneys. Small quantities are also made in the ovaries and brain. Low levels may be associated with infertility, while high levels may interfere with ovulation.

DHEA is an interesting hormone because it has limited functions on its own from what we know thus far; it mostly acts as a precursor to other hormones like testosterone and estrogen. It also serves as an indicator of adrenal gland function. Its effects are similar to those of testosterone, with contributions to the changes at puberty, energy, sex drive, bone health, menstrual cyclicity, fertility, and metabolism.

Thyroid Hormones

The thyroid gland sits in the neck, beneath the Adam's apple. It produces two hormones called thyroxine (T4) and triiodothyronine (T3). There's also TSH, thyroid-stimulating hormone, which is released by the pituitary gland in the brain.

The thyroid is responsible, along with other organs, for regulating metabolism, body temperature, energy storage and expenditure, hair and nail growth, bile production, digestive function, and–important for menstruators–ovulation and sex hormone release. The thyroid is also fairly sensitive to the body's physiological stress responses whether those are stemming from mental-emotional distress, infection, or low food intake. What this means is that thyroid function tests can naturally fluctuate, and repeat blood tests, after an out-of-range test, are fairly standard to understand what's going on.

Thyroid disorders, including the autoimmune kind, are prevalent in females and can lead to a variety of menstrual issues like irregular cycles, anovulatory cycles, heavy bleeding, scant bleeding, PMS, infertility (including early pregnancy loss), and even early menopause. Autoimmune thyroid conditions tend to be underdiagnosed due to the testing algorithms in place even though they're common. Thyroid testing is covered in Chapter 14.

Fun fact: the same hormone, kisspeptin, that contributes to gonadotropin-releasing hormone production is also involved in thyroid function. This is yet another overlapping connection between the thyroid and the menstrual cycle.

BOX 4. How does hormone regulation work?

The thyroid is a great gland to illustrate how the body knows when to make more or less of a certain hormone. Most of our hormones operate through a negative feedback loop. A negative hormone feedback loop is when the presence of adequate hormone signals the body to not make more of it. On the flip side, when there's not enough around, it signals to the body to make more.

When there are enough thyroid hormones (i.e., T4 and T3) being made, the brain detects this. Consequently, the brain decreases thyroid-stimulating hormone (TSH) production, which in turn decreases thyroid hormone (T4 and T3) production. On the contrary, when there's a demand for thyroid hormones, the body recognizes that and increases TSH production so that we can make more T4 and T3.

It's kind of like a thermostat regulating the temperature in a room. If you've set the thermostat to twenty degrees Celsius (that's sixty-eight degrees Fahrenheit) in the winter, then the heat will go on when the temperature starts to go below that point, or off when the temperature detected is at twenty degrees or more. If you're wondering, that is the temperature I like my thermostat at in the winter—it's a point of contention with my husband!

Cortisol

Cortisol is a hormone many of you have probably heard of. It is referred to as the stress hormone because it is a major player in the body's physiological stress response orchestrated by the HPA, hypothalamic-pituitary-adrenal, axis. Ovarian hormones, estrogen and progesterone, interact with this HPA axis.

Cortisol gets blamed for pretty much anything and everything, when it's simply trying to do its job. I get it, we live in a culture of hustling and grinding; chronic stress, burnout, metabolic conditions, lack of social infrastructure, climate anxiety, injustice, and busyness are rife. We definitely want to address these factors as much as possible to have hormone harmony (more on that in Part Five). However, we also have to understand the importance of cortisol and what it does for us.

Cortisol is crucial to survival and life. It's needed not only for your stress response but also for regulating your metabolism, getting you out of bed in the morning, making sure that your immune system is functioning well, regulating your mood, and controlling your inflammatory responses.

Like DHEA, it's produced in the adrenal glands, which sit on top of your kidneys. It is made from cholesterol.

Cortisol levels are typically high in the morning and low at night; this is part of your normal biological clock called the circadian rhythm. It naturally fluctuates throughout the day and is supposed to be made in higher quantities when you encounter a transient physical or mental-emotional stressor, like exercise, a presentation, not eating enough, a difficult conversation, someone chasing you, etc. It has a catabolic response in these situations so you have access to cellular energy, i.e., glucose, to react appropriately.

In the menstrual cycle, preliminary data indicates that we make more cortisol in general in the follicular phase; however, our cortisol response to stressors is lower. On the other hand, in the luteal phase, our baseline cortisol levels are lower; however, cortisol reactivity to a stressor is higher. This is pretty interesting as we start thinking about how menstruators tend to feel over the cycle: more reactive in the luteal phase, and with some fairly distressing diagnoses happening then as well (like PMS and PMDD).

When too high or too low, especially for prolonged periods of time for whatever reason, cortisol levels can lead to disturbances in the menstrual cycle, metabolism, thyroid, and immune systems directly and indirectly through its communication network. There seems to also be a difference between cortisol and HPA axis responses when an experience is adverse versus traumatic based on the severity of the experience. Age is another factor that comes into play. Then, there are congenital conditions as well as medications that can alter cortisol production and lead to menstrual (and other) disturbances. These are times we want to pay special attention to cortisol and have a plan of action in place.

Insulin

Insulin is made and released by the pancreas to help you utilize and regulate your blood sugar. It plays an important role in energy metabolism, testosterone production, appetite regulation, cortisol-mediated hormone responses, and more. It primarily has an anabolic effect, i.e., it's involved in storing energy. It is naturally higher in your bloodstream when you've eaten.

Our fat cells and liver cells have the most amount of insulin receptors, which are sites on our cells where insulin can bind and do its job.

The sensitivity of your cells to insulin is an important part of your metabolism, menstruation, and cardiovascular health. Conditions like type 2 diabetes and PCOS involve a loss in this sensitivity to an overproduction of insulin, called insulin resistance. Oftentimes, there is a genetic predisposition that causes alterations in insulin signalling; this may be exacerbated by lifestyle factors, our environments, and other structural issues. In turn, this can lead to exaggerated insulin production and other downstream effects such as a disruption in ovulation, sex hormone dysregulation, mood disturbance, weight change, inflammation, fatty liver disease, elevated cholesterol, and cortisol problems. All of this can directly and indirectly contribute to menstrual cycle disturbances.

There is a whole network of players involved in the menstrual cycle. Hormones aren't inherently good or bad–they are simply signals. They get their cues from players in different parts of the body and are influenced by the goings-on in life. Even though we know a lot about our hormones, there's still much to uncover.

As you go through different seasons and stages, there will likely be transient shifts as your body adapts. There will also be longer-term alterations as you move through your cycling years. The one thing that is guaranteed as you move from your first period to the last one ever is change.

CHAPTER HIGHLIGHTS

- Hormones are chemical messengers that tell your cells to do or not do something by binding to receptors on the cell surface.
- The menstrual cycle is governed by communication between the ovaries and two parts of the brain called the hypothalamus and pituitary gland. The trinity of these organs is called the hypothalamic-pituitary-ovarian (HPO) axis.
- GnRH (gonadotropin-releasing hormone) is released by the hypothalamus. It stimulates FSH (follicle-stimulating hormone) production from the pituitary gland to help grow and mature egg-containing follicles, and a spike in LH (luteinizing hormone) for ovulation to occur.
- The primary estrogen you make during your menstrual cycle years is estradiol. Estrogen rises in the follicular phase, dips at ovulation time, and rises again in the luteal phase. It thickens the endometrial lining, contributes to LH rising, and stimulates cervical mucus production. Estrogen gets a bad rap but it is crucial for health! Its other jobs include breast and mammary duct development during puberty and pregnancy, proliferation of vaginal and vulvar cells, and supporting bone metabolism.
- Progesterone is made after ovulation by the corpus luteum. Its primary job in the menstrual cycle is to mature and stabilize the endometrial lining. It works in opposition to and together with estrogen across the body's systems.
- Testosterone and DHEA (dehydroepiandrosterone) are the body's main androgens, i.e., hormones associated with traits belonging to males.

They are precursors to other sex hormones like estrogen. We need both of these hormones to thrive; when they are too low or too high, menstrual cycles, energy, fertility, libido, and metabolism can all be affected.

- The thyroid gland sits in the neck and makes thyroid hormones, T4 and T3. It helps in regulating metabolism, body temperature, energy balance, hair and nail growth, digestive function, and sex hormone release.
- Two other hormones that are involved in the menstrual cycle are cortisol, your stress hormone, and insulin, your blood sugar regulation hormone.

CHAPTER 5
From Menarche to Menopause

The menstrual cycle has been deemed a vital sign by the American College of Obstetricians and Gynecologists–the others are heart rate, respiratory rate, blood pressure, and temperature. This makes total sense because a menstrual cycle requires cooperation and collaboration between the brain, ovaries, and the rest of you.

Obviously, you could have a normal-looking menstrual cycle and there could be something brewing under the surface; however, having a menstrual cycle within the normal range indicates, for the most part, that:

- the brain and ovaries are communicating well
- hormones are being produced in adequate amounts in a timely fashion
- you have sufficient resources
- you're not under imminent danger and your body is able to adapt to day-to-day circumstances
- your organs are responding appropriately to your hormones

From the first period to the last, every menstruator has their own journey of ebb and flow. And it all starts with menarche.

Menarche

Menarche (pronounced *me-nar-key*) is when you get your first period ever, and it happens once the transition of puberty has begun. The average age of menarche is twelve years old, but it can happen sooner or later than that.

While in the womb, we have millions of follicles. Of these follicles 50 to 70 percent are lost around the time of birth, at which time one to two million egg-containing follicles are left in each ovary; these follicles are now called primordial follicles. Primordial follicles are in a state of arrested development until puberty is reached. The time between birth and puberty has been described as a neurobiological brake of reproductive function, which simply means that reproductive processes are on hold to allow time for growth and development.

As you go through puberty, secondary sex characteristics such as breasts and hair start to manifest. This all happens because the hypothalamic-pituitary-ovarian (HPO) axis has been activated, i.e., the hormonal communication between the brain and ovaries has begun. Once the HPO axis is switched on, hormonal changes lead to your primordial follicles becoming primary oocytes (immature eggs). A group of these oocytes can now grow with each cycle under the influence of follicle-stimulating hormone (FSH) and estrogen, with one of the egg-carrying follicles becoming the dominant follicle.

It's common to see some irregularities in cycle length and bleeding patterns in the first twelve or so months after the first period. A majority of menstrual cycles happen within twenty-one to forty-five days, but shorter or longer cycles may occur in the first year or so of your periods starting. Irregularities in the first year are usually not concerning. As the HPO axis begins to communicate in the menstrual way, the incidence of anovulatory cycles–cycles without an egg being released–is a bit higher. This is normal as the body learns its cyclical rhythm. It's like creating a new path through a forest: it's barely visible at first, but after being tread on many times, it gets clearer and clearer.

Irregular and anovulatory cycles may contribute to scant bleeds, heavier bleeding, and/or more painful periods. Then, there's stress, school, commitments, food and nutrition, genetics, medical history, and many other factors that tend to contribute as well. That being said, you don't have to suffer through symptoms like period pain. Getting some support in place, especially educational, lifestyle, and nutritional support, can be of great utility here (Chapter 15 has considerations for period pain). Plus, it *is* possible that there could be something else going on that warrants medical attention. Unfortunately, too many adolescents are dismissed, not assessed properly, or simply slip through the cracks due to their young age/ this is a huge disservice to young menstruators. Conditions like endometriosis can manifest early in one's life.

As of the last decade, more and more adolescents are reaching menarche earlier, even as early as nine years old. There's some genetic involvement here, but other factors such as early childhood growth, medical diagnoses, medication use, and exposure to hormone disruptors in and out of the womb are likely contributing factors. When secondary sex characteristics happen before the age of eight and/or menarche happens sooner than nine years old, it is called precocious puberty; specific and individualized care should be implemented to support short-term and long-term health. Precocious puberty is discussed in Chapter 21.

On the flip side, some menstruators get their first period much later than the average age of twelve years old. This, too, can be due to a variety of reasons such as genetics, medications, lifestyle factors, and medical conditions; however, the most common reason for a delay in menarche is due to a relative lack of energy or fuel to initiate and sustain menstrual activity. This is commonly found in high-level athletes, people with disordered eating patterns or eating disorders, those experiencing malnourishment or food scarcity, and those who have undergone reproductive organ surgeries. If you or a loved one hasn't had a period by age fourteen to fifteen, or have had your period disappear for months on end after showing up a few times only, please consult your doctor.

Menarche can be accompanied by excitement, fear, nervousness, apathy, and even dread–my experience was one of fear and shock as I shared in the Introduction to this book. When I noticed blood in the toilet for the first time,

I was in a panic! I knew nothing about periods at thirteen years old in Karachi and was almost in tears when I ran to my mum and told her about my bloody discovery.

As far as I remember, my education was: "*Tumhare periods shuru ho gaye hain,*" which translates to "Your periods have started." Not much of an explanation, but my mum didn't know what else to tell me apart from the fact that they happen monthly for a few days and that you need products to put in your underwear. I didn't make much of it at the time. I realize now that she, like many other menstruators, knew very little about periods and how to talk about them. It wasn't much of a conversation in her family, and just like me, she hadn't learnt about it in school. I didn't really learn about the menstrual cycle and why it happens or what it means until much later.

Some adolescents enter quietly into menarche, some are celebrated as they begin menstruating, some are afraid, some are welcomed into "womanhood," some are traumatized, some are relieved, and others feel confused and dysphoric as they realize that their physiology does not match who they are, and some, like me, are oblivious.

Everyone has their own story and experience. Regardless of what that is, I hope to encourage a bigger discussion around these topics in homes and schools, among adolescents and adults, and in public forums so we can spread awareness, learn from each other, have a better understanding of our bodies and the signals they send us, dispel some of the stigma that persists around this topic, seek timely care, and have doctors believe us when we are struggling or noticing that something is off.

Periods Through the Ages

Once your cycle is established, it should be fairly consistent from cycle to cycle with minor deviations from your norm. You can refer back to Chapter 3 to review what an average menstrual cycle looks like.

There might be times when aspects of your cycle change transiently, like delayed cycles when you're under a lot of emotional stress or if your energy needs aren't being met consistently; however, your typical pattern will probably return once you're through to the other side of that (perhaps with a little help).

Remember, the body allocates resources to preserving survival–not reproduction–under high stress, especially if it's prolonged. Everyone has a slightly different threshold as to what will alter a menstrual cycle and the kind of changes experienced. I have at least one delayed cycle during the winter if I've fallen ill with the flu or another upper respiratory tract infection before ovulation time.

Over time, you may notice an overall shift in cycle length, bleeding patterns, cervical mucus, and/or your premenstrual experience. These changes can happen for a variety of reasons. When you're younger, the likelihood of structural issues contributing to period issues is lower–nutrition status, energy availability, stress and trauma, medical history, and the establishment of a cyclical rhythm are the biggest contributors to period patterns. Just because an underlying condition is less likely, it does not mean we dismiss our younger patients.

As you age, the prevalence of structural issues like fibroids, polyps, and other conditions like thyroid disorders can have a bigger impact on menstrual cycles and how they are experienced. Different stressors, medical conditions, medication use, pregnancy history, past surgeries, self-awareness, cellular ageing, proximity to menopause, and the cumulative effects of your lived experience all play roles here. It's always prudent to bring up persistent and drastic deviations in your menstrual pattern to your healthcare team. It's also important that you bring up troublesome symptoms at any age so that you can be properly assessed and supported.

BOX 5. What about fertility?

Fertility education and age-appropriate conversations about fertility are severely lacking in healthcare. Even when there are medical conditions (cancer is one of the few exceptions), patients are not fully informed of the impacts on their fertility in the long run. This is a huge disservice to patients. It not only hinders time-sensitive decision-making but also causes a lot of distress.

Too often I get infertility and subfertility patients in my practice who are caught off-guard by how long they've been trying to conceive, who didn't fully realize the correlation between age and fertility, who received an oversimplified blanket diagnosis of fertility issues because of another diagnosis, who are surprised that their family doctors are referring them to a fertility clinic, who reach their mid and late thirties without a solid understanding of their reproductive health.

The term "infertility" was redefined in October 2023 by the American Society for Reproductive Medicine (ASRM) to be more inclusive and comprehensive. The updated definition (taken directly from the ASRM), describes it as: "a disease, condition, or status characterized by any of the following:

- The inability to achieve a successful pregnancy based on a patient's medical, sexual, and reproductive history, age, physical findings, diagnostic testing, or any combination of those factors.
- The need for medical intervention, including, but not limited to, the use of donor gametes or donor embryos to achieve a successful pregnancy either as an individual or with a partner.
- In patients having regular, unprotected intercourse and without any known etiology for either partner suggestive of impaired reproductive ability, evaluation should be initiated at 12 months when the female partner is under 35 years of age and at six months when the female partner is 35 years of age or older."[2]

---

[2] *American Society for Reproductive Medicine, Definition of infertility: a committee opinion, 2023.*

Fertility is a complex topic. Globally, we're seeing an increased incidence of male- and female-factor infertility. The average age of pregnancy has also changed.

As we age, our egg (and sperm) quality and ovarian reserve decline. From what we know, there is a mild change in egg quality from the twenties to thirties, a slightly bigger decrease from the early thirties to late thirties, and an even bigger reduction from the late thirties to forties on average. Not only that, the incidence of health conditions–reproductive and otherwise–also goes up with age. This, along with the cumulative effect of environmental exposures, lifestyle, and general ageing are all factors contributing to fertility and pregnancy outcomes.

Depending on future fertility goals and where you're at in life, freezing eggs and/or embryos is a prudent consideration earlier on, like in your 20s to early 30s. This doesn't mean that you can't freeze eggs later; however, the quality will be relatively lower and you likely will respond to hormone stimulation that's part of the procedure differently.

Earlier intervention for infertility procedures such as intrauterine insemination (IUI) and in vitro fertilization (IVF) impacts success rates positively. Unfortunately, these procedures are costly and not accessible to everyone. Even with private insurance plans, much of the fees cannot be recovered. In Ontario, the first round of IVF is covered by the provincial healthcare plan, but required medications, specialized testing, time off from work for the many appointments, and mental-emotional support are not.

Difficulty with conceiving and carrying to term are beyond the scope of this book, but if you have a condition like PCOS or endometriosis, there are considerations for you in Part Four of this book.

Conversations about fertility and reproductive health are important to have with yourself, your partner or family unit, and your healthcare team. You can find some relevant resources on these subjects in Appendix B.

Charting your cycle and using fertility awareness-based methods (FABMs) to pinpoint your fertile window and ovulation are great places to start (see Chapter 11 for more on that). Taking care of your overall health and keeping up with screening tests are always going to be important aspects of bolstering your fertility and pregnancy outcomes in the long run; Part Six of this book is all about that. There are antioxidants like coenzyme Q10 (CoQ10) that can be taken under the guidance of a healthcare professional for egg quality as well. A prenatal supplement with folic acid is the general recommendation for anyone of reproductive age who is sexually active and able to fall pregnant.

Pregnancy and Postpartum

During your reproductive years, there can be a break from the menstrual cycle in the form of pregnancy. If an egg is fertilized by a sperm, the fertilized ovum divides and is now called a blastocyst. This blastocyst will make human chorionic gonadotropin (hCG), which is detectable by at-home pregnancy tests as well as bloodwork. hCG, in turn, will signal to your corpus luteum to continue to secrete progesterone as progesterone is required for maintaining a pregnancy. The corpus luteum of pregnancy lasts until around weeks eleven to twelve of gestation, at which point it deteriorates. The placenta has been growing throughout this time, contributing to progesterone levels until it pretty much takes the progesterone reigns at the end of the first trimester.

If pregnancy happened but did not carry forward for some reason–pregnancy loss, ectopic pregnancy, or abortion–there can be a slight delay in your cycle starting back up, which is natural. Sometimes, cycle length, flow, and even period-related signs and symptoms can change.

Pregnancy tends to encompass a time frame of forty weeks, which means, in most cases, there are no periods for approximately nine months. After that, the return of the period can depend on factors such as pregnancy outcomes, lactation, hormonal changes, medical history, thyroid function, stress, and nutrition status. On average, the first period arrives at around the ten-week mark postpartum. One of the hormones that plays a role postpartum is prolactin, which is the lactation hormone.

Prolactin's main roles are milk production and the development of mammary glands within breast tissue. Prolactin is typically produced by the pituitary gland; however, during pregnancy, it is produced by the placenta as well. High prolactin levels can suppress ovulation and lead to irregular cycles. This is typical during lactation which can influence the return of the period depending on the frequency and duration of breastfeeding. It's important to note that you can have ovulatory cycles with minimal or no bleeding during lactation, so contraception should be used if a pregnancy is not desired.

The Menopause Transition

There's a lot of similar-but-slightly-different lingo out in the world concerning the time around the last period ever.

- The menopause transition refers to the years leading up to menopause.
- Menopause, on the other hand, is when your periods have been gone for twelve months. From The Menopause Society: "Menopause is defined as the final menstrual period and is usually confirmed when a woman has missed her period for twelve consecutive months (with no other obvious causes)."[3]
- Combine these two concepts and you get perimenopause. Perimenopause is essentially the menopause transition plus the one year until you officially reach menopause. "Peri" means around, "meno" refers to menstruation, and "pause" means stopping. So, perimenopause quite literally translates to the time around the cessation of menstruation.
- There's also post-menopause. I feel like this is the clearest! It's when you've left all of the definitions behind, i.e., you're beyond the first anniversary of your final period ever.

---

[3] *The Menopause Society, Menopause 101, 2024.*

The average age of menopause is between fifty to fifty-one years old. Early menopause is when menopause is reached between forty and forty-five years of age, while primary ovarian insufficiency (previously called premature ovarian failure, discussed in Chapter 21) is when menopause is reached before the age of forty. Then, there is surgical menopause which is when menopause is abruptly reached due to removal of the ovaries (with or without the uterus). The ovaries produce sex hormones, so even if you've had your uterus removed but still have at least one ovary, sex hormones will still be produced.

If menopause is reached early, there is strong evidence to initiate hormone replacement therapy (using estrogen and progesterone) to decrease the cardiovascular disease and bone loss risks associated with losing sex hormones at a younger age. If this is you, please consult with your healthcare team.

The age group between the late thirties and early forties often are under-supported and dismissed when it comes to hormone health because, technically, they're still in their reproductive years; however, we know that perimenopausal changes can begin many years (yup, even a decade) before we reach the end of our menstruating years. If you're in your late thirties or early forties, noticing shifts in not just your menstrual cycles, but also other parts of your health like sleep and mood, could indicate the beginnings of perimenopause. Signs and symptoms associated with the hormonal change can persist well beyond the final period ever.

The perimenopausal experience can vary vastly from person to person. Changes that surface around this time can impact fluid retention, appetite, energy, mood, metabolism, cholesterol, vaginal lubrication, incontinence, and memory. The more telltale signs of perimenopause include cycle length changes (of over seven days, usually closer in earlier stages and delayed in later stages), night sweats, hot flashes, flow changes, increased cramping, sleep changes, and/or body composition change, especially more visceral fat deposition around the abdominal organs. Some of these may carry over to after menopause and last for a few years, but not always.

Previously well-managed conditions might also have a resurgence and require a modification in strategy.

It's important to not simply chalk everything up to perimenopause–a thorough assessment of your concerns is always valid; some factors to consider would be thyroid disorders, insulin resistance, and nutrient deficiencies like iron, vitamin B12 and vitamin D.

We can't really capture and confirm perimenopause through blood work. Blood work may only confirm perimenopause as we get closer to menopause; FSH levels may increase in late perimenopause, but not so much in the early- to mid-stages. FSH remains high as the final period ever is reached, while estrogen and progesterone levels are low.

The incidence of anovulatory cycles usually increases in perimenopause and the production of progesterone is variable. It *is* possible to get pregnant during the menopause transition because all you need is one egg (and you can be in that transitional phase for years); however, it is less likely due to the reasons discussed in Box 5. Estrogen production can be a bit erratic, especially because we may have stacked cycles where a new cycle starts even though the last cycle didn't finish; consequently, we can have significant fluctuations in estrogen and our other hormones.

Ebbs and flows in both estrogen and progesterone play a role in how you experience your cycles during these years. We also can't forget that these hormones have a broader impact beyond our reproductive organs as discussed in Chapter 4, like cardiovascular health, metabolism, mood, bone density, muscle mass, and more.

All of that is to say that perimenopause is a time of significant transition, and everyone has their own experience through it. The changes you see in yourself are not only a direct result of shifting sex hormones but also the complex interplay of other biochemical processes in the body along with ageing. For some people, it's fairly smooth, and for others it's not. And don't forget: life also goes on simultaneously; whatever is happening in life feeds into your experience.

It probably sounds like a lot (and it can be!), but we can help smoothen the perimenopausal journey. Exercise (both resistance and cardio), diversified nutrition habits with plentiful plant foods, pleasure, good sleep, connection, mindfulness, access to robust healthcare, and having purpose all help create a solid foundation–the time to start is now, wherever you are on the menstrual timeline. We have tools like hormone therapy, medications, acupuncture, counselling, herbs, and supplements that can aid you along the way. Community groups and your support systems are crucial as well!

Menopause has been surrounded by stigma and taboo historically (no surprise there), with many servings of misinformation. This has prevented many folks from normalizing some of the natural changes around this time. It has also hindered medical support for troublesome symptoms and early-stage metabolic concerns while getting in the way of people living their lives as fully as possible.

The tides are changing and menopause is a hot topic in pop culture and the news right now. I've been really happy to see more and more myth-busting and media coverage on the topic. There's still a lot more work to be done to make sure we have age-appropriate and *accurate* information getting to the public, especially regarding the wide spectrum of perimenopausal experiences, therapeutic options for management and risk reduction, and embracing the changes that this time brings.

The lack of cycling hormones takes some getting used to. It tends to be bittersweet–it can be liberating to no longer have to think about periods, and it can be simultaneously saddening as one phase of life ends and a new phase begins. Appendix B has some books and resources on perimenopause and beyond.

CHAPTER HIGHLIGHTS

- Menarche is the first period ever, with an average age of twelve years old. It happens when the hypothalamic-pituitary-ovarian (HPO) axis is activated around puberty time after a neurobiological brake of reproductive function between birth and puberty. It's common to see some menstrual irregularities in the year after menarche as the body learns its cyclical rhythm. Factors that can impact period patterns include nutrition status, energy needs, stress, and medical history.

- Once a cycle is established, it should be fairly consistent from cycle to cycle with minor deviations. There might be subtle changes with time. Bigger changes can be due to structural issues, medical conditions, drugs, pregnancy and surgery, self-awareness, ageing, proximity to menopause, and the cumulative effects of your lived experience.
- If you have fertility goals, please remember that fertility is multifactorial and can change with age.
- Pregnancy is a time when there are no periods for around nine months, plus approximately another ten weeks postpartum. Prolactin, the lactation hormone, can intermittently interrupt ovulation in the postpartum months.
- The menopause transition refers to the years leading up to the final menstrual period (ever!). Once twelve months have passed since the final period, menopause is officially reached at the thirteenth month. Perimenopause is a term inclusive of the menopause transition and menopause, i.e., the years leading up to the final period plus the twelve months after. Once menopause is reached at the thirteenth month, the post-menopause era begins.
- Perimenopausal changes can impact cycle length, bleeding patterns, mood, temperature, body composition, and more. Fluctuations in estrogen and progesterone are part and parcel of this transition and can manifest in a variety of different ways. Support can include exercise, nutrition guidance, supplements and herbs, acupuncture, hormone therapy, medications, and community support.

PART TWO

TRACKING YOUR CYCLE

CHAPTER 6
The Importance of Cycle Tracking

Tracking my cycle was a game-changer as I tried to figure out what was going on with me. I started with the very basics: writing down when my period started and ended. This was in the days before period tracking apps were a thing, so good ole paper and pen did the job.

It's the first step you can take to learn more about your menstrual cycle and period. It's how I learned that my cycles would be extra-long around exam season. It's also how I learned just how irregular my cycles were in general.

Over time, I started to pay attention to the other things happening in my body and noted them alongside my menstrual cycle data. I noted down my energy levels, mood, acne, cervical fluid, and bowel movements as these were some of the things I was concerned about. I didn't know how to piece anything together, but I got to know my body a bit better and had some information to share with my healthcare team to answer the many questions I had.

The terms charting and tracking are often used interchangeably. I'm not sure if someone coined these terms; however, the difference seems to lie in the details and purpose.

The term charting appears to have come from fertility charting and actually using a daily table—aka chart—to note down ovulatory metrics like cervical mucus and basal body temperature, especially using some of the natural family planning trademarked methods covered in Chapter 11. On the other hand, tracking can be a bit more flexible in its scope, not necessarily geared toward fertility. Scientific literature doesn't particularly distinguish between the two, so I'm going to stick with the term tracking.

Tracking allows you to tap into your menstrual rhythm and get to know the nuances of your cycle. It isn't just about noting down the first day of your period (although it can be!); it's about getting to know your body throughout your whole cycle. You don't know what you don't know.

As hormones shift, and you move from one part of your cycle to the next, there will be differences–some subtle and some not so much–and only you can do the work to figure out what's going on. Some reasons to start tracking your cycles and associated signs and symptoms are as follows:

- Helps you calculate your cycle length
- Offers insights into your cycle and body, providing a gauge of your overall health
- Helps identify any areas of concern (and if everything is A-OK!)
- Provides pregnancy prevention and fertility data
- Helps you be specific in your communication with your healthcare team
- Aids in self-advocacy so you can seek out support for your cycle and symptoms
- Enables you to pick up on subtle changes over time
- Allows you to lean into your menstrual framework where your cycles support you, and you support your cycles
- Provides a long-term record of your menstrual cycles

This last point is especially important because our memories aren't set in stone. In fact, our brains can edit and update our memories based on new information, evolving values, and other variables. What this means is that sometimes we may remember, favour, omit, and ascribe more meaning to certain information over others–yup, we're biased! On top of that: we forget things!

I'm not sure about you, but I've hidden items around the house in a seemingly clever spot, thinking how could I *ever* forget that ... just to turn the house upside down trying to find the mystery location later. So, take a page out of my book, and note things down.

Similar to keeping track of blood work and blood pressure over the long run, your tracking data can offer valuable insight into progress and patterns over time. For example, if you tracked longer cycles or delayed periods any time you had exams like me or when you had a major project coming up, then you know that that's your inclination. Or, you might notice that your cycle length remains constant, but your mood swings and fluid retention might be off the charts in these kinds of stressful circumstances. We're all a bit different and not everyone has the same menstrual cycle.

As covered in Chapter 5, periods will change over time as well. If you're curious about something, are noticing major deviations from your normal experience, are having cyclical or troublesome symptoms, or are alarmed because of something you read online, jot your questions down and discuss them with your healthcare team at your next appointment.

Menstrual cycles are dynamic, responding to your life and environment. As such, some transient changes as the body adapts are natural. With tracking, more is not always better. The goal is to notice trends over time. Hypervigilance is not the point; overanalyzing every sign, symptom, or shift can be a bit of a slippery slope into a downward spiral of anxiety!

If you're wondering what you should be tracking, then stay tuned as the next chapter has the details.

CHAPTER HIGHLIGHTS

- Tracking your cycle is the first step you can take to get to know your body and cycle.
- Cycle tracking allows you to identify changes from your norm, pinpoint your fertile window for pregnancy or contraception, communicate with your healthcare team, and have long-term data on your health.

- The goal with tracking is to notice trends, and not to hyperfocus on one-off changes.

CHAPTER 7

The Ins and Outs of Cycle Tracking

This chapter is a breakdown of what and how to track your cycle. For those of you new to this practice, I encourage you to begin with a few basics like menstrual flow and any changes occurring before your period starts. Then, you have the option of layering in more detail.

An easy-to-remember acronym is POOP:

- Period
- Ovulation signs and symptoms
- (Optional) Overall and other, and
- Premenstrual experience

Period

This P is all about your flow. When you're tracking your period, consider jotting down answers to some of the following questions:

- *When does it start?* This will automatically allow you to calculate cycle length as well.
- *How many days does my flow last?* Remember: a period begins with the first full day of flow. If there's spotting for a couple of days before the period, then that goes under premenstrual experience.
- *How much blood am I losing?* This can be challenging to calculate because most menstruators change their products before they are

entirely saturated if using an absorbent product, or completely full if a cup or disc is being used. It can be even more difficult to calculate if multiple or homemade products are being used and if there's some free flow or flooding. Most products should give an approximate amount of fluid that they can hold. The average amount of blood loss is thirty to fifty millilitres per period, where ten to seventy millilitres is considered a normal amount of blood loss. It's important to note down any significant and persistent changes in your blood loss volume even if they fall within range. For example, if your blood loss doubles from thirty millilitres to sixty millilitres, it's prudent to discuss this with a healthcare provider even though this new number technically falls within the normal range of blood volume loss. This is of particular importance if other changes are happening simultaneously (like pain or clots).

- *What is the consistency and colour of the flow?* As discussed earlier, the colour of menstrual blood can be different shades of red with some small clots being fairly common; however, large and/or numerous clots warrant an investigation.
- *What other signs and symptoms are present with each period?* The things to think about here are energy levels, mood changes, pain levels, pain location, painkiller use and other management methods, bowel movements, etc. Identifying these will allow you to discuss symptoms with your healthcare team, diagnose any potential conditions, and put appropriate measures in place if required. Some of you might simply record the huge sense of relief as severe premenstrual symptoms resolve with the onset of a period. Note it all down!

Ovulation Signs

Ovulation can bring along a variety of signs and symptoms. These are discussed in detail in Chapter 3 and may include cervical mucus changes, pain in your lower abdomen called *mittelschmerz*, change in BBT (basal body temperature) right after ovulation, and even increased libido. Some of you might be quite sensitive to ovulatory hormone changes and may notice symptoms such as headaches and breast tenderness, or have an underlying condition that might lead to mid-cycle spotting.

Optional: Overall and Other

You can paint this O with whichever colour you'd like to. The way I used this myself and in my practice is to record the change in a particular symptom or health concern that is new, that comes and goes, or that is important to patients. Fluid retention, fatigue, acne, and headaches commonly fall under this category.

This allows us to not only identify patterns and whether there is a hormonal relationship, but also leads to great conversations about the non-hormonal variables involved.

For example, fluid retention is commonly influenced by progesterone, but also by salt and potassium intake, heavy exercise, and dietary carbohydrates; energy levels can be affected by hormone shifts as well as iron and vitamin B12 deficiency, sleep, food and exercise habits, stress, etc.; headaches, especially migraines, can be impacted by changes in estrogen, but also by dehydration, irregular eating, alcohol, coffee, muscle tension, sleep, and more.

Ultimately, tracking your symptom of choice alongside your cycle can make it convenient to track said symptom, help normalize it if it's what we expect to see, and facilitate diagnosis or treatment. It could also show that there's no relationship to hormonal fluctuations at all, which can point us in the right direction. It can offer clarity.

Premenstrual Experience

Some mild changes before the onset of a period are natural, like changes in fluid dynamics and appetite, as estrogen and progesterone levels shift. Just knowing that should offer some relief (I hope!). They might be more noticeable if you're experiencing high stress, poor sleep, and inadequate, low, or irregular food intake.

If you feel like your life is being affected significantly by your hormones and where you are in your menstrual cycle, you might be onto something. Tracking the premenstrual experience can help discern if something is only happening in the luteal phase, like premenstrual syndrome (PMS).

PMS can show up as breast tenderness, headaches and migraines, sleep changes, irritability, weepiness, anxiety, low mood, acne, constipation, bloating, spotting, abdominal pain, back pain, and cravings that influence your quality of life. PMS doesn't have to *pretty much suck* (notice what I did there?!). There are numerous options available to you that are covered in Chapter 16.

Tracking can also help us understand if you have a premenstrual exacerbation (PME) of an existing condition. This is when you have a disorder that gets worse in the luteal phase and goes back to baseline soon after the onset of period flow. Let's say you've been diagnosed with a depressive disorder. With every cycle, you notice that the depressive symptoms intensify significantly after ovulation. You also recognize that things improve once your period starts. This might be due to PME, which is a kind of premenstrual disorder. PMDD (premenstrual dysphoric disorder) also falls in this category. For more on these diagnoses, see Chapter 17.

So, the best way forward is to record your signs and symptoms over a few cycles. Pay attention to the duration, intensity, frequency, and impact on your daily functioning.

Ultimately, we all respond differently to hormonal fluctuations, so sometimes tracking a certain symptom yields nothing, sometimes it provides information to investigate further, and sometimes it simply opens up further discussion.

How often should I track?

Tracking your cycle is a practice that takes time, effort, and mental capacity just like anything else in life. Since the goal is to determine your patterns, you can't simply go on one month's worth of data. You want to track for at least a few cycles–three or more—fairly comprehensively. Since periods change and life happens, there might be times when you want to take extra notes. The menopause transition is a good example of when it can be helpful for you to be more diligent and regular with your tracking.

After you've done your baseline tracking of a few months to understand your cycle, you can become a bit more flexible in your approach. Perhaps, you simply record when you get your period. Maybe add in any significant changes from your baseline.

My tracking practices have changed over the years: I used to record a lot more detail in the past, but now my approach is a bit looser as my cycles are regular and I am in tune with my body a lot more. I track the first and last days of my period, I'll note down any cervical mucus, and then I'll only note down things that are uncommon for me or old symptoms that have returned if and when they occur.

It has never been easier to record your cycles. Thanks to advances in technology and apps, it has become super easy to record your cycles and the signs and symptoms that accompany them. Certain apps sell your data, so that might be a consideration for you (some apps that don't sell data are listed in Appendix B).

An app is a piece of technology: it can make predictions based on algorithms that use the data available to it from other users and you–it does *not* replace body literacy. Some of my fertility patients discover in our appointment that their app's ovulation prediction is off (which obviously will impact their efforts to try to conceive).

You're the expert on the goings-on in your body, and tracking can help you understand them a bit better. The goal is *not* to be hypervigilant or hyper-reactive. When we're anxious or stressed out, there's a higher chance of misunderstanding or misinterpreting what the body is telling us. It's best to not jump to conclusions when you're starting to pay attention to your cycles, and simply gather some information.

The idea is that you learn about your body, familiarizing yourself with its tendencies, identifying anything that's amiss or getting in the way of you living your life fully, notice any changes happening over time, and address the questions and/or abnormal findings with your healthcare team.

Perhaps, it'll allow you to pace yourself a bit better throughout the month, or realize that your night-time cravings for any and all snacks leading up to your period is actually your body signalling that you need more food throughout the day (more on cyclical living and nutrition in Part Six of the book).

You're the only one who can do this work to figure out your patterns and tendencies. So, pull out your phone, use your calendar, write in a journal, create a spreadsheet, doodle and draw–track in a way that's easy and/or fun for you.

CHAPTER HIGHLIGHTS

- Tracking your cycle doesn't have to be super complicated. Note things down in a way that is easy for you: an app, calendar, journal, spreadsheet, etc.
- A good place to start is with the period itself, noting down when your period starts, when the flow stops, the quantity of blood, any accompanying symptoms and sensations, etc.
- You can use the acronym POP or POOP to guide you:
    - Period
    - Ovulation signs and symptoms
    - Overall and other (optional)
    - Premenstrual experience
- To figure out the intricacies and trends of your body, you'll need a few months of data. And remember, the goal is not to overanalyze but to identify overarching patterns so you can get more in tune with your body and seek out care if and when needed!

PART THREE

CONTRACEPTION

CHAPTER 8
Contraception Considerations

Contraception has had a tumultuous history across the globe. In Canada, it was illegal to advertise or sell birth control until 1969, at which point contraception was (finally!) decriminalized so that all Canadians could have the right to it. Abortions, under extremely restricted circumstances for life-threatening pregnancies, were also decriminalized at this time.

Today, there are many districts and nations where there is little to no access to these healthcare measures. In Canada, we theoretically have universal healthcare, but this healthcare is inequitable. We have access to contraception, abortion, and fertility care, but not everyone has the same access.

Reproductive healthcare and justice aren't only about access to contraception and abortion, but also about access to healthcare providers, sexual and reproductive health education, timely diagnosis and shared medical decision-making, practical and affordable treatment strategies, food and financial security, affordable safe housing, transit and transportation infrastructure, freedom from discrimination, and other structural and systemic factors. We'll return to some of these elements in Part Five of this book.

We must continue to work toward improved access to dignified and universal healthcare for *everyone*. Just in the last couple of years, we've seen some steps taken backward to criminalize access to safe abortion care as well as censorship of sexual and reproductive education in a few parts of the world. It shouldn't need saying, but everyone deserves to have agency and autonomy over their bodies and medical choices.

Raising awareness and engaging in conversations with your families, friends, and communities is a good first step. Joining advocacy groups, writing to local and federal politicians (or getting involved!), donating your money and time to these efforts while not supporting those undermining basic rights and freedoms, and amplifying existing platforms are all great ways to contribute.

Historically, females have had to carry the burden and responsibility of contraception, even though menstruators are only fertile for approximately a week per month. The only two common contraceptive measures for males have been physical barriers like condoms and a vasectomy, which is a surgical procedure.

According to Dr. D. J. Handelsman, PhD a couple of years back:

> Despite the proof of principle, strong community interest and medical priority on the need for new, reversible male contraceptives, pharmaceutical industry development has ceased. This market failure reflects pharmaceutical industry perceptions of low profitability coupled with high risk in the context where better return on investment is expected for developing drugs in more lucrative, less risky areas of chronic disease compared with the higher product liability risk of developing novel contraceptives for healthy men to compete with existing low-cost female methods.[4]

I can't say it any better. It's a skewed game—a numbers game, for the most part. We see this same narrow-minded, capitalist, and patriarchal narrative play out in many areas of health and healthcare.

Thankfully, menstruators are getting involved, medical guidelines and practice are evolving, and the tides of research are turning. There are still numerous cracks (and gaping holes) to fill and ample room for growth. At the same time, there's lots of change afoot. Menstrual research is at an all-time high with more funding and studies on the menstrual cycle, hormonal fluctuations, reproductive physiology, hormone therapy, and natural solutions. There's even

---

[4] *Handelsman, Endotext [Internet], 2019.*

a (teensy) bit of emerging research on novel male contraceptives. Furthermore, we're studying more diverse populations than ever before.

Contraceptives like the pill and IUD have been around for a while. They've decreased the number of unwanted pregnancies, offered better control over one's fertility, and reduced the need for unsafe abortion. Their non-contraceptive benefits have helped millions of people with their menstrual and hormone-based symptoms. They have changed lives, helping menstruators progress leaps and bounds since their advent, providing time and space for education, career-building, and whatever the heck you want to do; this includes offering an effective choice to opt out of having kids altogether.

I'm not saying these tools are ideal solutions, especially for menstrual and hormonal conditions for which they are used frequently. Hormonal contraceptives receive a lot of criticism in the wellness space, and rightfully so.

The combined oral contraceptive pill works by shutting down ovulation and, thus, the menstrual cycle. This affects the whole body. The pill is often prescribed for a plethora of hormone-based symptoms, sometimes without a comprehensive investigation to diagnose or understand those symptoms. Moreover, the nature of the treatment is typically not discussed adequately. Prescribing doctors need to do a better job of ensuring they're having risk-benefit conversations to facilitate informed decision-making. The pill is covered in detail in Chapter 9.

That being said, the pill has a vast body of evidence to support its use for a variety of conditions while preventing unwanted pregnancies. It can offer peace of mind with respect to pregnancy prevention for better mental and emotional well-being; this is especially true if you lack menstrual education and the capacity or resources to seek out additional and alternative care. On top of that, it's relatively inexpensive, and even subsidized by the government in many countries.

Cost is often a hindrance to accessing effective birth control; it disproportionately affects low-income and younger people. Private insurance may cover or offset these costs; however, not everyone has access to that. If you live in a place without universal healthcare, then getting a prescription may be associated with a cost as well. Choice and access are important aspects of informed decision-making. Too often costs get in the way of supporting one's health.

Where I live, only people under twenty-five years of age could access hormonal birth control including certain contraceptive pills and hormonal intrauterine devices at no cost under the provincial healthcare plan called OHIP (Ontario Health Insurance Plan) if you had a prescription from your healthcare provider. Just recently, in 2024, the Canadian government announced the roll-out of a national universal pharmacare plan that includes contraceptive care and diabetes meds. This is a huge step and a tremendous win for anyone wanting and/or needing contraception. Here are typical costs per patient associated with common contraceptives[5]:

- Oral birth control pills: up to 300 Canadian dollars per year
- Hormonal intrauterine system: up to 500 Canadian dollars per unit (units may last for three to seven years)
- Copper intrauterine device: up to 100 Canadian dollars per unit (units may last up to ten years)
- Hormonal implant: up to 300 Canadian dollars per unit
- Hormonal vaginal ring: up to 300 Canadian dollars per year
- Medroxyprogesterone acetate injection: up to 150 Canadian dollars per year
- Levonorgestrel emergency contraceptives: approximately 30 Canadian dollars per dose

[5] *Freeland Chrystia Deputy Prime Minister of Canada, Government highlights next step to universal access to free contraceptives, 2024.*

These are the categories included in this universal access to no-cost contraceptives. It's not an exhaustive list, but it is quite comprehensive, which is great!

This has happened after concerted advocacy efforts over time (much gratitude to everyone and anyone involved!). This is the kind of pressure we need to continue to put on our local, provincial, and federal politicians. This kind of infrastructure results in healthcare system savings.

Ultimately, there's no singular way to do contraception or health. It's *your* choice to make. It has to fit where you're currently at. You also get to change your mind as your health, life, values, body, experience, finances, knowledge, and circumstances change over time.

There is–and should be–no shame or judgment in accessing care that meets your needs. Asking for care should not be an act of resistance. Accessing care should not require you to cross borders. These are things that shouldn't need to be said, but here we are in the age of petty lawmakers trying to snatch away bodily autonomy and pointless online commentary and trolling by complete strangers.

Everyone deserves to have access to care that meets their needs; not only some people.

CHAPTER HIGHLIGHTS

- Contraception has had a tumultuous history, and bodily autonomy continues to be under attack in many parts of the world.
- When deciding to use any kind of contraception, a thorough conversation including how it works, pros and cons, and efficacy should all be discussed.
- If you're using hormonal contraception for non-contraceptive benefits, then a work-up and comprehensive assessment should be performed to understand the nature of your symptoms and if you have a diagnosis of a long-term complex condition. This is what patient-centred care with informed decision-making looks like.

- There is no singular “correct” way to do contraception. It is your choice and you can change your mind at any time.

CHAPTER 9
The Pill

The oral contraceptive pill, or birth control pill, is the most commonly used hormone therapy globally. The hormones used in these pills are like the estrogen and progesterone you make in your body but with slight variations. They are non-bioidentical, i.e., they are not exactly the same as the hormones your body makes.

There are numerous kinds of pills today with different forms, doses, and combinations of hormones. The kind of estrogen used in many pills is ethinyl estradiol. For this book, "the pill" refers to the combined oral contraceptive pill, which has estrogen and progestin both, and we'll call the kind of pill that contains progestin on its own the "progestin-only pill". Progestins are progesterone-like compounds that differ structurally from the progesterone we make but act similarly to progesterone in many ways, especially in the uterus; they can have varying other effects–sometimes even opposite or foreign to progesterone–in other parts of the body.

How does the pill work?

The pill essentially blocks your body's menstrual cycle by inhibiting ovulation. It contains estrogen and progestin. This suppresses the production of your brain hormone, gonadotropin-releasing hormone (GnRH).

The way the pill does this is through a negative feedback loop by telling your brain that there are enough hormones around and that there's no need to make more. Consequently, your luteinizing hormone (LH) is suppressed, which means you won't be able to ovulate. Theoretically, the pill will shut down follicle-stimulating hormone (FSH) production as well; however, it doesn't always do that.

Progestin-only pills are usually prescribed for those who want to avoid the combination pill, like when there is a high risk of developing clots and strokes, or if someone cannot use and tolerate estrogen therapies. They can also be used for contraception postpartum, although intrauterine devices are gaining popularity in that population.

Progestins might suppress the secretion of your brain hormones involved in the menstrual cycle; however, their primary action is to make the endometrium less favourable for implantation by keeping it thin. They can also thicken cervical mucus. So, even though fertilization is possible–at least half of users ovulate while on a progestin-only pill–the chances of implantation are next to nil.

The hormones in oral contraceptive pills have changed over time, with many more options available today. Different pills will have varying combinations, dosages, and types of hormones. As such, side effects, benefits, and tolerability can vary from pill to pill. A deeper look into the numerous combinations and types of estrogens and progestins found in the various birth control pills available today is beyond the scope of this book. There are a few resources in Appendix B you can refer to for that.

The kind a doctor prescribes to a patient depends on what's going on and what the goal is. Are you going on it for contraception, to suppress ovulation because of PMDD (premenstrual dysphoric disorder, discussed in Chapter 17), to decrease testosterone and acne, to have less pain and heavy bleeding, or for a few different reasons? Even when the goal is clear, a bit of trial and error with a few different pills might be needed depending on tolerability and symptomatic relief–not everyone responds to the same things in the same way!

Some people are prescribed extended-use pills that have no placebo weeks or will have a seven-day placebo week every three months. Most people, however, will be put on oral contraceptive pills in a way where there are twenty-one days of hormones followed by seven days of placebo pills or no pills during which no hormones are used. The drop in hormones during the placebo week triggers a bleed to occur. This makes a full (monthly) round of the pill look like a menstrual cycle.

This is not a menstrual cycle, though, and the flow you get while on the pill is called a pill bleed or withdrawal bleed. Taking away the progestin destabilizes any lining that had built up, and leads to flow. It is similar to what happens in the body: you shed your endometrial lining when progesterone drops. The two are not the same because the natural cyclical hormonal rhythms and ovulation are missing when you're on the pill.

BOX 6. Is the pill linked to fertility problems?

The short answer: is no.

If you went on the pill for a non-contraceptive reason, like irregular menstrual cycles, cystic acne, heavy flow, or painful periods, then those could be linked to infertility depending on the cause of those symptoms.

You can also have new developments that impact your fertility while you're on the pill—not necessarily as a consequence of use, but simply because you were on it for a while and your health changed during that time.

You may simply not have a good idea of what's happening in your natural menstrual cycle if you've been on it for a while since the pill shuts your cycle down. It's when you discontinue it that your old or new symptoms show up.

For example, let's say you were prescribed the pill for menstrual cramping and heavy bleeding, the cause of which was not determined and perhaps even chalked up to you being young and athletic.

Once you start on the pill, you find that those symptoms are now non-issues (hurray!). Since you bleed every month and your flow is much lighter, you think that your period problems are behind you.

Years later, you decide to come off the pill because you are thinking about fertility. You may find that your period woes of yore have resolved, and all is well in the cyclical world of your hormones.

However, you could also see a return of your pain and heavy bleeding when you go off the pill. Moreover, you may notice other developments like mid-cycle spotting or a worsening of your old symptoms compared to your pre-pill days. This is especially true if you had or have an undiagnosed condition that was causing your symptoms.

Getting a consistent dose of hormones was managing your symptoms by suppressing the ups and downs of your personal hormone production whilst you were on the pill. When you no longer have that treatment in place, whatever the underlying cause was (like endometriosis), is still there; endometriosis is associated with fertility issues. More on that in Chapter 20.

There is a marked difference between knowing all of this information and making the decision to go on the pill because it's the best treatment option for you at the time (which is what we want: informed decision-making!) versus being told that the pill will regulate and lighten your periods without digging deeper and thinking about the bigger picture.

This latter situation in which there was inadequate investigation and explanation potentially delays diagnosis, may lead to suboptimal treatment, can impact future fertility and family planning efforts, and might even lead to regret for not trying harder or advocating more.

Pros, Cons, and Contraindications

All oral contraceptive pills are approximately 99 percent effective if you use them perfectly. "Perfect" use refers to taking the pill exactly the same way every day as is done in a research setting. Typical use, on the other hand, is how most people use the pill in the real world (where you switch up times and miss some pills every so often). With typical use, the failure rate is 9 percent, meaning it is effective at preventing pregnancy 91 percent of the time.

Now, let's go over some pros of using the pill, which will vary slightly depending on the combination of hormones you're on and who you are:

- Effective method of contraception when used properly
- Allows for the cessation of menstruation or provides some control over it
- Helps decrease heavy and prolonged menstrual bleeding
- Helps decrease menstrual pain
- Can help with hormonal mood disturbances and physical symptoms like those seen in PMS and premenstrual disorders
- Decreases the risk of ovarian and endometrial cancers
- May decrease ovarian cyst formation in some people
- Can help with excess androgen symptoms like acne and hirsutism by lowering free testosterone
- Can help with endometriosis symptoms
- Correlated with decreased risk of developing bacterial vaginosis
- Is often inexpensive (averaging around twenty Canadian dollars per month) and even subsidized by the government in some places. Canada has announced a national universal pharmacare plan in 2024 that includes no-cost contraception

And, now, let's go through the cons:

- Have to remember to take it daily
- Does not offer protection against sexually transmitted infections (STIs)
- May decrease libido and lead to vaginal dryness
- Is associated with an increased incidence of vaginal yeast infections
- Linked to worse mood, especially depression but also anxiety and suicidal ideation. The risk of depressive symptoms increases if you have a history of depression
- May increase the need for nutrients like vitamin B6, zinc, and magnesium
- Increases the risk of liver, brain, and breast cancers

- Impacts brain chemistry, thyroid function, and the coordinated stress response between the brain and adrenal glands
- Can worsen acne in some people and is associated with weight gain (directly and indirectly through mood and energy changes)
- Contributes to elevated inflammatory markers and disturbed gut microbiome
- Can decrease peak bone mass if started in adolescence
- Long-term costs add up if you don't have insurance or subsidization programs where you live
- Increases the risk of developing blood clots and cardiovascular issues (high cholesterol, stroke, increased inflammatory markers, etc.). This risk of clots and strokes is very slim. If you are someone who has been impacted by this, please remember that this wasn't your fault! The unfortunate reality of the pill (and many treatments) is that the benefits are discussed but the risks aren't discussed enough, especially when the outcome is rare. Even when protocol is followed, we can't always predict who will be affected in this way. Genetic factors or other unknowns are almost impossible to know. The latest research on young people who have had strokes while on the pill is pointing to the involvement of genomic factors, i.e., your genes and how they interact with one another and the environment, that put you at higher risk for early-onset stroke. I'm hoping we'll gain much-needed clarity around this soon.

The Society of Obstetricians and Gynecologists of Canada (SOGC) and the World Health Organization (WHO) both concur that there are absolute contraindications to using the pill–the combo pill. An absolute contraindication means that the risks outweigh the benefits and the treatment should not be used. These are listed below:

- Being over thirty-five years old while smoking fifteen or more cigarettes per day
- A current diagnosis of cancer

- Hypertension (i.e., high blood pressure) with a measurement of 160 mmHg systolic (the top number of a blood pressure reading) or 100 mmHg diastolic (the bottom number of your reading), or higher
- Being less than six weeks postpartum
- Having migraines with aura
- History of stroke or ischaemic heart disease (where there is an impediment to blood flow to the heart)
- Having multiple risk factors for cardiovascular disease–think diabetes, high cholesterol, high blood pressure, etc.
- Past or current venous thromboembolism (i.e., a clot forming in a vein) like deep vein thrombosis or pulmonary embolism
- Prolonged immobilization due to surgery and/or injury
- Known thrombogenic mutations that put you at higher risk for developing clots
- Liver cirrhosis, liver tumours, and/or acute hepatitis
- Diabetes of over twenty years and/or diabetes with vascular complications
- Systemic lupus erythematosus (SLE) with positive or unknown anti-phospholipid antibodies (APA)

Then, there are relative contraindications, meaning that caution should be exercised, the pros and cons should be weighed thoroughly before the pill is used, and robust monitoring should be undertaken as it's being used. These relative contraindications include:

- Being over thirty-five years old and smoking *less* than fifteen cigarettes per day
- Breast cancer that has been recovered from for five or more years
- Hypertension that is controlled with medication or where the blood pressure measurements are between 140 to 159 mmHg systolic (top reading) and/or 90 to 99 mmHg diastolic (bottom reading)
- Dyslipidemia, i.e., cholesterol problems
- Symptomatic gallbladder disease or history of gallbladder issues with the use of the pill

- Using certain medications like rifampicin, rifabutin, and anticonvulsant drugs
- Migraine without aura

The progestin-only pill is an option for those who are at higher risk of cardiovascular disease and clotting, cannot use estrogen, or have a history of migraines with aura.

You should be able to find all this information publicly—most national medical associations will have guidelines based on the current available data in different age groups and populations. You can find links to the SOGC and WHO guidelines in Appendix B.

Using the Pill

Originally, the pill was used in populations over twenty years of age for family planning purposes. Over the last few decades, the use of the pill in adolescents has increased. This has allowed for fewer unwanted pregnancies that can change the trajectory of someone's life, especially considering the state of abortion care in some areas. It has also helped with decreased heavy bleeding, cramping, and acne that affect school, work, sports, other commitments, and even self-esteem.

Putting someone on the pill without discussion, especially for non-contraceptive purposes in adolescence, is a lost opportunity to investigate these patients and educate patients about lifestyle, nutrition, stress, and how those impact our bodies and menstrual cycles.

Not eating enough, not getting adequate calcium and other minerals, lack of exercise, overexercising, and stress all have real impacts on periods, especially period pain and PMS. Starting with those basics, or combining them with the use of the pill, is a much better option than simply offering medication without a discussion.

Starting the pill young is also associated with a slight decrease in peak bone mass, i.e., the maximum amount of bone mass you can accumulate (ever!), and we're starting to see some data on it being linked to brain chemistry alterations. This makes sense considering hormonal interplay across the body.

Hormonal contraceptive pills have come a long way. But general support for menstruating people has not. I don't particularly like how the pill is commonly used: like a crude science and medicine.

Instead of figuring out why someone's having irregular cycles or heavy bleeding, and what foundational measures might be missing, the solution is to prescribe the pill to shut down the system entirely. It's kind of like that one closet or room in the house we keep throwing stuff into, and instead of cleaning and organizing it, we shut the door so that we can't see it anymore.

Don't get me wrong: sometimes we may need, want, or choose to do that, and that is okay! It can be an effective and cost-effective management option.

The thing is: that hormones are messengers. They don't work in isolation; the pill influences not just your sex hormones, uterus, and ovaries, but the rest of you as well. Your nutrient status, mood, libido, digestion, and everything else will be impacted because most tissues have receptors for sex hormones.

The pill is used by millions of people globally, and there's a lot of data on it. It is a tool that is available to you on your menstrual journey. It's a tool that has prevented pregnancies, offered relief from symptoms, provided peace of mind, and helped menstruators live their lives.

It's also a tool that is not appropriate for everyone and may cause harm to some.

You can ask your doctor to review the prescription and take some baseline data that is part of routine testing and screening like blood pressure, cholesterol, liver function, etc. If you're using it for non-contraceptive purposes, then imaging, testing, and examination relevant to you and your case should be performed. And, obviously, you can take your own notes on how you're feeling, like mood, energy, and libido. This way, if there are new developments after you start the pill, you can bring these up, while monitoring any improvements you're experiencing as well. This can be applied to starting any new medication, supplement, or other treatment regimen.

Pretty much all health decisions come with risks and benefits; this includes doing nothing for contraception and/or menstrual symptoms. Ultimately, all you can do is make the best decision you can with the information you have and then reassess regularly to make sure it's still serving you.

Even though I cannot prescribe the pill due to my scope of practice in Ontario, I regularly have conversations with patients about their past, current, and/or potential use of hormonal contraceptives. It is part of my job to lay out their options, discuss risks and benefits, and facilitate decision-making.

If you're thinking about going on or continuing with the pill, I encourage you to have a frank discussion with your healthcare team so you can feel confident in whatever decision you make!

CHAPTER HIGHLIGHTS

- The combined oral contraceptive pill has non-bioidentical forms of estrogen and progesterone to prevent pregnancy. It shuts down ovulation, and, hence, the menstrual cycle. The bleed experienced on the pill is a hormone withdrawal bleed.
- The pill is often used for reasons beyond contraception like heavy menstrual bleeding, PMS, PCOS, and period pain.
- The pill is frequently used as a crude science to treat hormone-based conditions, sometimes without further investigation.
- It isn't associated with fertility; however, it can hinder the detection of hormonal and menstrual issues while it is being used as it often treats and manages the manifestations of those very concerns.
- Progestin-only pills are used in people who are at a higher risk for cardiovascular disease and stroke, or who cannot tolerate estrogen.
- The risks and benefits of using hormonal contraception aren't discussed enough, and the risks are sometimes minimized. It's always good practice to review any treatment regimen with your healthcare team so that you're making decisions that are in line with your current and future goals, taking into account relevant personal and family history.
- Remember, health decision-making is complex; there's no one-size-fits-all solution. Even doing nothing comes with risks, especially with

respect to pregnancy prevention. Your choice of contraception to prevent unwanted pregnancy is one that you should make with your medical team. You can reassess at any given point.

CHAPTER 10

Intrauterine Contraceptives

Intrauterine contraceptives (IUCs) are one of the most effective ways to prevent pregnancy–they have an overall failure rate of less than 1 percent in the first year of use (which is fantastic!). They are small instruments that are inserted into the uterus. There are non-hormonal and hormonal kinds available. You might've heard of the copper intrauterine device (IUD) and the hormonal IUD or IUS, which stands for intrauterine system.

The use of intrauterine contraceptives has gone up globally since they were first offered for contraception. As of 2019, depending on the country, anywhere from 2 to 40 percent of menstruators use some form of intrauterine contraceptive.

Intrauterine contraceptives have come a long way since their conception. They didn't have the T shape that we're accustomed to today as seen in Diagram 9.

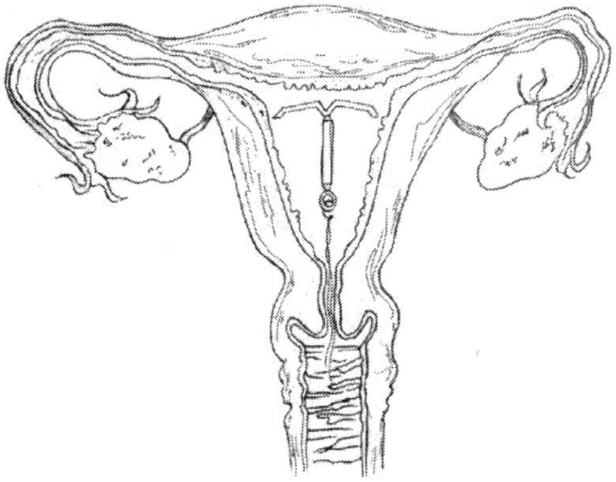

DIAGRAM 9: *T-shaped intrauterine device (IUD) in the uterine cavity.*

Other shapes that have been found around the world include spheres and frameless products, which come in hormonal and non-hormonal varieties. We're going to stick to covering the most commonly used T-shaped devices here.

Intrauterine contraceptives have a high up-front cost, depending on where you live and what resources you have access to. The copper IUD is around 100 Canadian dollars, while the hormonal kind can be anywhere from 150 to 500 Canadian dollars. There might be additional costs associated with insertion and any associated check-ups.

Insertion of an intrauterine contraceptive is a medical procedure. Your doctor will conduct a vaginal and cervical examination using a speculum, insert the device into the uterus, and remove any extra insertional bits. Removal of the intrauterine contraceptive is also a medical procedure, where your doctor will use a tool to retrieve the device from the uterus through the cervix.

Insertion of these devices can be quite painful and is an under-discussed part of the process. The internet is riddled with experiences from people with significant pain and trauma from device insertion. The usual advice of: "Just take some ibuprofen prophylactically" is not helpful; research shows that ibuprofen prophylaxis is insufficient at offering pain relief for IUD insertion. There are painkillers and local anesthetics like lidocaine that are available. If you ever decide to get an IUD, please talk to your doctor about the procedure before insertion, discussing pain relief options as well.

Intrauterine contraceptives have become popular as they don't carry the same cardiovascular risks and other side effects that the birth control pill does.

IUCs have an extremely low rate of failure, which basically means that they are excellent at preventing pregnancy. On the rare occasion that a pregnancy does occur, however, there is a high chance of it being an ectopic one (i.e., an unviable pregnancy where a fertilized egg implants outside the main cavity of the uterus). This risk is higher with the hormonal one versus the copper IUD.

Intrauterine contraceptives are contraindicated during pregnancy, in the presence of local malignancies, and certain anatomical or structural issues. There is a very small risk of uterine perforation regardless of whether they're hormonal or not. This chance is slightly higher if inserted while breastfeeding or less than thirty-six weeks postpartum, especially if someone is both breastfeeding and less than thirty-six weeks postpartum. The risk is highest in the first two to three months after insertion and happens in 0.3 to 2 percent of these cases.

Other adverse effects include developing a pelvic infection called pelvic inflammatory disease, which doesn't happen from the device itself but may be due to an existing infection like an STI (sexually transmitted infection).

Now that we've broadly discussed intrauterine contraceptives as a whole, let's get a bit more granular and review the commonly-used non-hormonal and hormonal devices.

Copper Intrauterine Device (IUD)

The copper IUD is a T-shaped plastic IUD with copper wiring and collars around the T-shaped frame. It's the only non-hormonal IUD approved by the FDA (Food and Drug Administration, in the USA) and can be used for up to approximately ten years, although it has shown efficacy for up to fifteen years.

The copper IUD increases the release of inflammatory compounds, leading to inflammation in the uterus. Copper ions from the copper IUD impair sperm motility and viability. All of this ends up making the uterus less hospitable to a potential fertilized egg.

This kind of IUD does not impair ovulation, meaning that you continue to release eggs and have a menstrual cycle.

It's also an excellent form of emergency contraception if used within five days of intercourse or ovulation, with a failure rate of less than 1 in 1000; this is a smaller failure rate than other forms of emergency contraception, which is covered in Chapter 13.

It's a great option for people who do not want to use hormonal medications for contraception; however, it might not be suitable for those of you who experience significant period pain or heavy bleeding. There is an increased chance of heavier menstrual flow and lower abdominal pain especially in the first three to six months after insertion–this is the top reason for IUD removal in 10 to 13 percent of users. It's also not suitable for those of you with a copper allergy.

Some risks, with respect to ectopic pregnancy and uterine perforation, were discussed earlier in this chapter. Other than that, you may also absorb a tiny amount of copper, which thus far hasn't been shown to be problematic. There is a small risk of the IUD breaking upon removal, and there might be some pain with intercourse. As always, discuss anything you're noticing in *your* body with your healthcare team.

Hormonal Intrauterine System (IUS)

The hormonal IUS, also called hormonal IUD, has been around since the 1970s and became popular once the Mirena IUS was introduced in the early 2000s. The IUS contains the progestin called levonorgestrel. It can be used for painful periods and heavy bleeding as well as contraception.

Its lifespan is shorter than the copper IUD with the recommendation to remove or replace it every seven years (this constitutes a recent change from the five-year replacement timeline that was used prior to the 2020s).

The hormonal IUS, like the copper IUD, is also a T-shaped device. Instead of copper wiring, it has a reservoir of the progestin levonorgestrel that gets released over time.

The Mirena, Liletta, and Levosert devices contain fifty-two milligrams of levonorgestrel and release around twenty micrograms daily for the first few months. By the end of year one, that dose is closer to eighteen micrograms per day. As the end of the device's lifespan approaches, that daily dose slowly declines to around ten micrograms.

There are other lower-dose IUSs like Skyla/Jaydess which release less levonorgestrel daily. This also means that they need to be replaced more frequently.

The levonorgestrel IUS works in a few different ways. It causes atrophy of the uterine lining so that it remains thin, and increases the production of a protein called glycodelin A, which interferes with implantation and thickens cervical mucus, which can impair sperm travel.

It can also lead to ovulation suppression; however, ovulatory cycles continue in most cases. The percentages vary depending on the study and population you're looking at, but anywhere from 20 to 50 percent of users may stop ovulating entirely, with the majority of that happening in the first year of use when the daily dose of levonorgestrel released is the highest. This also means that it won't help with PMS like the pill does.

Light bleeding is common with the hormonal IUS. This bleeding can be in accordance with a menstrual cycle, or it might be random, especially in the first few months after insertion. Approximately 20 percent of IUS users achieve a loss of period flow. The most common effect associated with the IUS is amenorrhea, i.e., the complete absence of bleeding or spotting for ninety days consecutively. It's a bit of a polarizing effect–some people love it, while others don't particularly like the fact that they no longer bleed.

A systematic review and meta-analysis in 2019 compiled data from multiple different studies and showed that the IUS increasingly led to amenorrhea over the span of a year—0.2 percent of users in the first ninety days, 8.1 percent in the second ninety days, 11.2 percent in the third ninety days, and 16.8 percent by the end of the year.[6] Proportionally, the biggest jump was seen between the first and second ninety-day intervals.

Now, let's get into some of the complications that can arise from an IUS. It's rare for an IUS to get out of place; it happens in 0.1 percent of cases, i.e., one in one thousand times, and it typically happens if there was some uterine

---

[6] *Sergison et al, Am J Obstet Gynecol. 2019.*

perforation as it was inserted or if there's scar tissue present. If it gets dislodged, it can end up in a variety of places, with the most common location being the abdominal cavity. In the rare case that a pregnancy does occur while a hormonal IUS is in place, there is an approximately 50 percent chance of it being an ectopic pregnancy. In contrast, the copper IUD carries a 6 to 9 percent increased risk. These numbers may seem high; however, remember that these are very effective contraceptive devices, so the chances of getting pregnant while you have one inserted are rare to begin with.

Overall, the hormonal IUS is an effective and safe option for a lot of menstruators. You're still getting hormonal contraception; however, the hormone is in small doses and the local delivery method (and potential for adverse effects) is different from taking hormones orally. It also doesn't contribute to cardiovascular risk like the pill. It can provide a cost-effective alternative to surgery for people who experience abnormal uterine bleeding, especially if surgery is not an option for whatever reason.

It can decrease the incidence of heavy menstrual bleeding significantly. It has been shown to decrease measured blood loss by 71 percent within six months and up to 90 percent after twelve months, while also improving quality of life. When heavy menstrual bleeding is decreased, it also means that any concomitant anaemia, missed work or school days, and associated costs like period products are also addressed. Furthermore, it can be a viable option for those of you with period pain and heavy menstrual bleeding due to adenomyosis as well as fibroids, which are discussed in Chapter 21.

It can also be used for endometrial protection from hyperplasia (uncontrolled and abnormal cell growth) induced by the medication Tamoxifen, typically prescribed to breast cancer patients.

Some people don't tolerate it well, as is the case with any treatment modality. As such, it may not be suitable for everyone. Systemic absorption of levonorgestrel through an IUS isn't particularly well-studied, nor are systemic effects (in fact, the data we currently have is sparse and inconclusive); they do happen, though.

There are numerous case studies, nascent research, and anecdotal evidence of people feeling unwell–most commonly with mood alterations and depression–after insertion of the hormonal IUD. I've seen this in clinical practice as well. Simply because a side effect is rare, it doesn't mean that we shouldn't warn patients or screen for mood changes. If you feel unwell, please bring that up with your medical team.

We know that ovulation suppression may happen in some individuals, especially soon after insertion. We also know that levonorgestrel binds to sex and stress hormone receptors when taken orally, which can lead to physical and mental-emotional symptoms. At the same time, we know that the amount of progestin being released by the IUS is really small and not the same as oral use. But ... we also know that we're all different. What this means is that we can't predict how someone will take to it. Even though most users tolerate it quite well, it's best to keep notes on how you're feeling so you can discuss any concerning changes with your doctors and move forward accordingly.

CHAPTER HIGHLIGHTS

- An intrauterine contraceptive (IUC) is inserted into the uterine cavity. There are hormonal and non-hormonal types available, typically shaped like the letter T. They are extremely effective as contraceptives.
- Insertion and removal of intrauterine contraceptives are medical procedures. The biggest risks of using these are uterine perforation and ectopic pregnancy, both of which are rare. Pain management during insertion especially is an often-neglected area of care.
- The copper IUD is a T-shaped plastic non-hormonal IUD with copper wiring. It does not affect ovulation. This device impairs sperm motility and viability and creates a local inflammatory response in the uterus. It makes for an excellent emergency contraceptive if inserted within five days of unprotected intercourse or ovulation.
- The copper IUD can worsen heavy bleeding and/or period pain, especially in the first three to six months of insertion.
- The hormonal IUS contains a reservoir of the progestin levonorgestrel, which is released in small doses over its lifespan of seven years. Most

users continue to ovulate on the hormonal IUS; however, a small percentage of users may stop ovulating in the first year of use.

- The levonorgestrel IUS works by thinning the uterine lining. It is used for contraception as well as heavy bleeding and period pain. It can significantly reduce menstrual bleeds to spotting or even amenorrhea (i.e., no bleeding).

CHAPTER 11

Body Literacy and Fertility Awareness-Based Methods

Getting a pulse on your hormonal rhythms is time well spent. Cultivating an awareness of the signals your body is giving you can be extremely helpful not only for fertility or pregnancy prevention but also for your overall health (and life!).

The fertile window is the timeframe during which conception is possible. Fertility awareness-based methods (FABMs) essentially draw on signs of ovulation for a non-hormonal and non-invasive approach to pinpointing your fertile window. There are many techniques and they all rely on charting your cycles.

A combination approach of using two or more methods offers the highest rate of accuracy in determining the fertile window. These are the same signs that many menstruators track for fertility purposes to time intercourse in an attempt to conceive. For contraception, the goal is to identify the fertile window and abstain or use barrier protection when the chances to conceive are highest.

One of the fertility awareness-based methods relies on cervical mucus charting. During the fertile window, cervical mucus production increases in response to rising estrogen. This mucus is typically more alkaline and slippery, creating a welcoming environment for sperm. Once progesterone production begins post-ovulation, cervical mucus tends to become thicker and tackier, which indicates that ovulation has already occurred.

Another method is the temperature method to figure out when there's an increase in basal body temperature (BBT). This temperature rise happens *after* ovulation when progesterone is released. This doesn't help as a prospective tool for the current cycle, but tracking this over time can indicate which day(s) of your cycle you tend to ovulate on. The temperature change is slight (approximately 0.3 to 0.5 degrees Celsius or 0.5 to 1.0 degrees Fahrenheit). A thermometer with two decimal places should be used first thing in the morning to take your temperature throughout your whole cycle. Keep in mind that BBT can be unreliable in certain instances like if you have a thyroid condition, are on certain meds, or do shift work. It is not recommended as a stand-alone method for this reason.

The calendar method, which relies on tracking Cycle Day One over six months, is based on using your cycle length to calculate your fertile window. You do this by subtracting nineteen from the total number of days in your shortest cycle; this forms the first day of your fertile window. You then find the longest cycle and subtract eleven from that number; this forms the final day of your fertile window. For example, if you recorded cycle lengths that varied from twenty-nine to thirty-two days, then the fertile window is between cycle days ten to twenty-one. There are other simplified methods similar to this one, but the cycle length methods shouldn't be used on their own.

Other techniques include using urinary hormones like measuring luteinizing hormone via ovulation predictor kits (OPKs), softer texture and higher/shorter cervical position, potential manifestations of ovulation like mittelschmerz, and fertility apps. Some fertility awareness-based methods have trademarked or registered trademarked names including the mucus-based methods of Billings and Creighton, calendar-based Standard Days Method, and urinary hormone Marquette Method (there are many others). Appendix B has more resources on these techniques, including one that synthesizes all the methods with their efficacies and costs while offering a shared decision-making tool for clinical use.

The trademarked methods, when employed individually, have a failure rate of 1 to 5 percent with perfect use; that rate becomes 2 to 17 percent with typical use. It's difficult to gauge rates with typical use when different combinations of FABMs are in place as we don't have that data yet, but the general recommendation from the literature and medical professionals is to combine methods for better success.

Preliminary research seems to indicate that there's a 1 to 2 percent increase in pregnancy rates when barrier protection is used typically (not perfectly) during the fertile window versus complete abstinence in that time, i.e., abstaining during the fertile window yields lower failure rates than using barrier protection. This makes sense because now we're kind of gauging success based on the failure rate of the barrier method.

With perfect use, that combination of barrier contraception plus FABMs is around 99 percent.

Long-term charting–at least six months–is needed to identify trends and decrease the chances of failure. Since menstrual cycles can present differently for everyone, the use of multiple methods is recommended in the literature.

FABM-based contraception success is directly linked to effort, diligence, and understanding. Education and receiving feedback from a trained educator or professional are key aspects of utilizing these methods, whether using them for contraception, conception, or even working on improving menstrual health conditions.

Many menstruators find this way of life empowering because they become more in tune with their bodies and it improves body literacy. With adequate guidance and when used as part of a shared decision-making process with medical professionals, it leads to improved outcomes and patient satisfaction. Small studies out of the USA have shown that while many physicians believe that fertility awareness methods are effective for patients wanting to achieve pregnancy (seems logical!), their disposition toward them is less positive regarding contraception. One of the challenges is that many physicians are unfamiliar with FABMs (including their efficacy), which impacts patient care. Family planning outcomes are more successful when multiple visits and education are offered. This requires resources and time.

Time, unfortunately (especially in the public healthcare system here in Canada), is a limiting factor when it comes to medical appointments. Resources, thankfully, have become prevalent online. There are dedicated low- and no-cost classes offered by many local and community organizations as well as by companies with trademarked methods.

FABMs may not be effective or practical in adolescence when cycles and patterns are being established, in perimenopause when cyclical changes can be erratic, for people on certain medications, when ovulation signs are not noticeable, and/or if you have unmanaged hormonal conditions that affect menstrual cycles (there's a few of those!).

Instead of DIY-ing it, it's best to speak to a practitioner knowledgeable in menstrual cycles and FABMs to ensure you're gathering and using your charting data appropriately. Unwanted pregnancy can have many consequences. As you learn, employing a different or additional form of contraception for pregnancy prevention is prudent depending on your situation.

CHAPTER HIGHLIGHTS

- Fertility awareness-based methods for contraception are ways to determine your fertile window by tracking changes around ovulation time and then abstaining or using barrier protection during that window to prevent pregnancy.
- These methods include:
    - Cervical mucus that is stretchy and clear
    - Temperature rising after ovulation
    - Using the calendar method of cycle tracking
    - Other methods like ovulation predictor kits, cervical position, and other manifestations of ovulatory hormonal changes
- A combination approach is most effective, using at least two, but ideally more, methods. Tracking should be done over at least six months to have better accuracy.

- These methods can be inaccurate and/or unreliable if your cycles are irregular, you're on certain medications, you're lactating, you've recently started having periods, or you have a diagnosis that affects ovulation or the signs tracked.

CHAPTER 12
Other Contraceptives

Depending on where you live and what regulations are in place, there are different kinds of products available to you for contraception and hormonal symptoms.

Barrier Methods

Condoms are the most commonly used barrier contraception. Condoms have a 2 percent fail rate when used perfectly (i.e., they are 98 percent effective at preventing pregnancy if used perfectly). With typical use, the failure rate is closer to 13 percent. They help prevent sexually transmitted infection (STI) transmission, can be found in a variety of different sizes and materials, and are the one barrier protection with options for both male and female anatomy.

Condom use, combined with the fertility awareness-based methods of temperature and cervical mucus charting, may provide around 99 percent protection against unwanted pregnancy when used *perfectly*. That rate does decrease with typical use by 1 or 2 percent compared to abstinence and fertility awareness. So, the combination approach might be a good option for those of you looking for contraception options while having a natural menstrual cycle, but it requires knowledge and diligence as discussed in Chapter 11.

Other barrier methods include diaphragms and cervical caps. Both are flexible dome-like silicone products. They cover the cervix and are used with spermicide. Basically, they both provide a physical and a chemical barrier to sperm. Neither is widely available in Canada. The failure rate of a diaphragm is

6 percent with perfect use and 12 percent with typical use, whereas the cervical cap has a failure rate of 16 to 30 percent.

The last common barrier method is the sponge, which is a two-inch-wide foam disc that contains spermicide. It is inserted into the vaginal canal to release spermicide and block the opening to the cervix. It has a little loop attached for removal. With proper use, it has a failure rate of 12 percent in the first year.

If you're using lubricant with a barrier method, please make sure that your lubricant is compatible with the product you're using. For example, oil-based lubricants should not be used with latex condoms as they impact the integrity of the latex. Lubricants are discussed in Box 7.

BOX 7. To lube or not to lube?

There are a lot of misconceptions about lubrication and what using a lubricant means. The stigma and misconceptions include those around texture, smell, messiness, and the implications that there's something wrong or that the reproductive pathways are "shrivelling up." Yup, that's what I hear in my office and out in the world on the internet and TV.

Lubrication is influenced by myriad factors including (but not limited to) arousal, your natural lubrication patterns, hormone levels, foreplay, where you are in your cycle, mental state, and medication use. Simply put: there's nothing wrong with using lube.

Lubrication, similar to libido, waxes and wanes. Using lube can facilitate a smoother and more pleasurable experience. It does not mean there's something wrong with you, although it's a good idea to talk to your healthcare team if you're noticing persistent changes or have questions about what a normal variation is versus something that requires attention.

Vaginal lubrication and female libido often go unsupported compared to erectile dysfunction, so this is an area where you might need to advocate significantly depending on your doctor.

More people than you think use lube and their lives are better for it. It can offer a more enjoyable experience with less pain and discomfort. If you don't like a particular lubricant, try a different one. You might need to try a few to find one that works well for you and your circumstance (water-based, oil-based, silicone-based, fertility-safe, etc.).

Hormonal Implants, Injections, and Patches

There are a few hormone-based contraceptives outside of the birth control pill and hormonal IUS.

One of these is the etonogestrel subdermal implant (Nexplanon), which is a small rod the size of a matchstick that's inserted into the upper arm just under the skin by a doctor. Subdermal implants have only recently become available in Canada.

The implant releases the progestin called etonogestrel similar to the levonorgestrel in the hormonal IUS. It can be left in place for three years and can lead to irregular bleeding patterns. These irregular bleeds lead to a high discontinuation rate in the first year of use.

Another progestin-based option is Depo-Provera, which is a shot containing the progestin called medroxyprogesterone acetate that is injected every three months. It has a failure rate of 6 percent and some side effects include irregular bleeding, spotting, headaches, weight gain, mood changes, dizziness, bloating, loss of libido, and local bone loss (in the area of the shot).

Then, there's the patch, which is kind of like a nicotine patch that is placed over the areas of the body that naturally have more fatty tissue like the belly, buttocks, upper arm, or back. It releases estrogen and progestin transdermally–i.e., through the skin–and must be changed weekly and worn for three consecutive weeks. Then, a no-patch week is usually implemented. It's like taking the pill without having to take the pill, and the effects are similar to the pill as well. It has a failure rate of around 9 percent.

Nuvaring is next: it is a vaginal ring that contains estrogen and progestin, which is inserted into the vagina for three weeks and then removed for a week. It's about 91 percent effective and has similar effects to the pill. The difference is that it does not carry the same increased risk of cancer; however, it can lead to increased body hair growth and irregular bleeding.

There are pros and cons to everything! You have to weigh those out and decide on a course of action with the guidance of your healthcare providers.

Surgical Procedures

If you're looking for a permanent contraceptive solution, then tubal ligation is an option for you. Tubal ligation is also known as getting your tubes tied. It can be done in a variety of ways: banding, cauterization, clipping, or tying and cutting. The main contraindication to tubal ligation is wanting to fall pregnant at any time in the future. Other reasons to not qualify for this procedure would be if you're not a good candidate for surgery.

There are places in the world where it's difficult to get a tubal ligation, and there are many reasons why you may be denied it, especially if you're young and don't have kids; there are some areas where you can't get one without a husband's permission even if you're not partnered ... that, folks, is the patriarchy at its finest.

If you are going forward with tubal ligation, make sure to discuss the procedure thoroughly with your doctor, as some people do experience regret after they've done it. There is a risk of failure, it does not protect against STIs, and it *is* a surgical procedure (typically done laparoscopically).

Depending on how the tubal ligation was done, there could be a reversal possible, and pregnancy can even be achieved with in vitro fertilization (IVF) or an anastomosis, which is where a connection to the ovary is created surgically, but these are costly procedures and do carry many risks.

Apart from tubal ligation, there can be surgical removal of the fallopian tubes, ovaries, and/or uterus. You might be familiar with the term, hysterectomy, which is the removal of the uterus. Something to remember here is that if the uterus is removed while the ovaries are left intact, there will still be estrogen and progesterone released by the ovaries.

Even though the study of gynaecology has a dark history of organ removal without consent for sterilization and exploratory purposes, organ removal is not standard practice. It is typically reserved for those who have other clinical indications for surgery like large intrusive fibroids or uterine cancer.

CHAPTER HIGHLIGHTS:

- Condoms are the most commonly used barrier methods with a failure rate of 2 percent when used perfectly. Combined with multiple fertility-awareness methods, they can be up to 99 percent effective at pregnancy prevention when used perfectly. They also help prevent STI transmission.
- Other barrier methods include the diaphragm, cervical cap, and sponge.
- Lubricants should always be checked for compatibility with the type of barrier protection being used. Lubricants can offer a more pleasurable experience regardless of age and barrier protection use, as internal lubrication can vary based on a multitude of factors.
- Hormonal contraception options beyond the pill and IUS include subdermal implants containing etonogestrel, Depo-Provera shots containing medroxyprogesterone acetate, a patch containing estrogen and progestin, and vaginal rings like the Nuva-ring containing estrogen and progestin.
- Tubal ligation is a surgical procedure that can involve banding, cauterization, clipping, or cutting the fallopian tubes.

CHAPTER 13

Emergency Contraception

It would be remiss of me to not include emergency contraception since it is more widely used than you might think. Anyone who needs it should have free access to it. It is an important element of reproductive justice and harm reduction. Emergency contraception is contraception to prevent an unplanned pregnancy in an instance where there was:

- no birth control used
- the use of contraception was not perfect
- a contraceptive method failure

In Canada, the most common emergency contraceptive pills are available over the counter but oftentimes kept behind the counter. What this means is that you might have to ask for them even though you do not need a prescription to get them.

In Canada and most other countries, Plan B, which is the progestin levonorgestrel, is the most common "day-after" pill that's used. It's also called My Way, After Pill, and Next Choice. It works by preventing or delaying ovulation. It's most effective if taken within twenty-four hours of intercourse but can be used up to seventy-two hours afterward, with the efficacy going down over time. That's why it's best to take it as soon as possible. There's some contention as to whether body weight or body mass index (BMI) might factor into its efficacy based on a couple of small studies that could use an update. The consensus among professional organizations like The Society of Obstetricians and Gynaecologists of Canada (SOGC) and The Canadian Pharmacists Association (CPA) is that you should take it regardless of size if that's all you have access to. You should not be denied it based on that either (unfortunately, that does happen). According to Health Canada, ulipristal acetate might be a better option for you if your body weight is over seventy-five kilograms (165 pounds) or your BMI is over 25 kilograms per metre squared.

The second pill available in Canada is ulipristal acetate (UPA), which goes by Ella. It is a bit more effective over a longer time frame of up to five days post-intercourse. The problem with Ella is that it typically requires a prescription from a physician, which can be difficult to get on short notice. There are systemic effects like nausea, cramping, dizziness, headaches, breast tenderness, heavy bleeding, and menstrual cycle disruption that you may notice with these pills; however, these effects are typically short-term.

Emergency contraceptives can cause slight changes in your next bleed. If your period is significantly delayed, it's prudent to confirm that pregnancy did not occur with a blood pregnancy test measuring hCG.

The most effective emergency contraceptive is one we've already gone over–the copper IUD. It can be inserted up to seven days after the incident by a doctor to provide contraception. But obviously, this requires a doctor who will be able to do this for you promptly in the first place, which is extremely unlikely.

Your local or national public health organizations should have a list of emergency contraceptives available to you where you live, how much they cost, and how to access them. In Canada, these medications cost anywhere from around fifteen to fifty Canadian dollars. Under the 2024 no-cost contraception plan, the copper IUD and levonorgestrel will be covered; however, UPA will not. Appendix B contains some resources that you can access for this information.

CHAPTER HIGHLIGHTS

- Emergency contraception helps prevent an unplanned pregnancy in instances where there was no birth control used, use of contraception was not perfect, or if there was a contraceptive method failure.
- The most common emergency contraception is Plan B, which is the progestin levonorgestrel which is available over the counter. A medication called ulipristal acetate may also be used but this usually requires a prescription.
- The most effective emergency contraception available is the copper IUD; however, a prompt doctor's appointment is required to insert this.

PART FOUR

COMMON MENSTRUAL CONCERNS

CHAPTER 14

General Menstrual and Hormone Health

The simplest way to get a pulse on your menstrual health is to start tracking your cycles. The best way to improve your general hormone health is actually to focus on the things that are supportive of your overall health like sleep, stress, nutrition, exercise, joy, purpose, connection, community, and routine testing; these pillars of health and wellness are covered in detail in the final few chapters of this book.

We're all a little different

A healthy menstrual experience can look a bit different from person to person. A menstrual or hormone-based condition can look different from person to person as well.

This is a great place to take a pause and think about what your goals are with respect to your hormone and menstrual health. Are you working toward general wellness? Are you hoping to prevent a condition that runs in your family? Are you experiencing distressing symptoms? Do you have a diagnosis?

If you have a menstrual or hormone-based condition, you likely need more targeted support beyond general recommendations or preventive measures. There is a difference between managing chronic non-communicable diseases versus working on improving well-being or working on prevention. They're not mutually exclusive, but they are different.

For example, someone who has mild discomfort with a period may find that ginger tea starting a few days before their period helps them take care of that. Someone with fibroids or adenomyosis may find that the same ginger tea barely touches their symptoms of pain. You can still have ginger tea in the second case, especially if you like it or if it helps, but it's markedly different than the first scenario.

Factors that can skew your cycles include sleep disturbances, stress, trauma, infections, non-menstrual health conditions, medication use, exercise, nutrient deficiencies, energy availability and food intake, body composition changes, and more. This might show up as a lack of or delay in ovulation (and hence delayed missed or delayed periods), more symptomatic cycles, or shifts in bleeding patterns.

Not all deviations need intervention; sometimes, regular programming resumes with the passage of time. Sometimes, a little extra TLC is required.

We all have different thresholds of what our bodies can tolerate and at what point things are going to shift. You and I could be going through similar situations but have vastly different experiences. On the flip side, you and I could be going through completely different things in life, but have similar manifestations in our health.

I know that my period is often delayed when I'm very stressed. This is frequently the case when I travel for more than a couple of weeks at a time. These trips require a fair bit of coordination, extra work, and stress prior to leaving. There's often a time change involved and my routine is different. Obviously, this will depend on where I am in my cycle. If I've already ovulated before the trip, then I'm going to get my period on time, approximately two weeks after ovulation. However, stress can delay ovulation and make my follicular phase longer. This automatically means that my period is going to come later as well. I know that my cycle will return to its usual rhythm afterward without me intervening.

I'm sure many of you can relate to this or think back to times of stress and change when your cyclical patterns changed (perhaps during the COVID-19 pandemic lockdowns? Or when you've suffered a great loss?). Tracking your cycles really is a good way to figure things like this out.

Your health doesn't exist in a bubble. It's a complex web of genetics, development, *in utero* (in the womb) exposures, upbringing, lifestyle, culture, society, privilege, healthcare, politics, resource availability, and much more. Your menstrual cycle and your relationship with it have been and will continue to be informed by these dynamic factors. Luck has a role to play as well (although sometimes "luck" is influenced by power, money, and privilege).

This isn't to say that you can't change the trajectory of your health–you can do a fair bit! We have significant influence over our health as discussed in Part Six of this book which lays out practical action items to consider even in the face of systemic barriers covered in Part Five.

It simply means that we have to take more of a bird's eye view and hold all of these realities together:

- your history and present circumstances contribute significantly to your health
- there are areas of health you can influence significantly
- there are some areas pertaining to your health and healthcare that are beyond your control
- there are many things about health we do not yet know

We do the best we can with what we know at any given time. Your best will probably be different than my best.

Keep this in mind the next time you compare yourself to someone who doesn't ever get acne, constipation, or "fill in the blank", while you're someone who tries hard and still has to deal with these. Keep this in mind when you compare yourself with someone who has a lot more resources (time, money, etc.) to spend on their health and well-being. Keep this in mind as you compare yourself to someone who has never experienced all the things that you may have experienced in life.

Even if we all did the same things in the same way at the same time, we'd all have varied experiences through all of that and end up in different places.

To Test or Not to Test

Sex hormone testing is a common topic that comes up to assess general hormone health as well as menstrual concerns. Don't shoot the messenger, but sex hormone testing for general wellness doesn't usually give us data we can utilize. Even for hormone-based concerns, sex hormone testing isn't always relevant. I'm not saying that it's always inappropriate to test sex hormones—I'm simply stating that it's not always appropriate.

For example, in cases of PMS or PMDD, we know that hormone changes are contributing to symptoms, but no diagnostic hormone test will tell us that you have PMS or PMDD. What you experience before your period is what helps us diagnose you. On the other hand, testing your sex hormones if your periods have gone missing while not pregnant furthers our understanding of the cause.

As you already know, hormone levels fluctuate over the menstrual cycle. We cannot draw a straight line from a certain hormone level to a symptom most of the time. You and I could have the same signs and symptoms but very different test results for estrogen and progesterone at the same spot in our menstrual cycles with controls in place for age and current lifestyle. Conversely, you and I could have the same test results for hormones but have very different signs and symptoms—heck, one of us could be feeling fantastic while the other feels crummy in this situation.

Sex hormones are only part of the picture, even in hormone-based conditions. Returning to the example of PMS and PMDD, even though sex hormone testing isn't really relevant, vitamin D testing is, as a deficiency is linked to the severity of those conditions.

The timing of sex hormone testing often needs to be considered as well. Low progesterone before ovulation is normal because progesterone production rises *after* ovulation. So, if you test low on progesterone on the ninth day of a thirty-one-day menstrual cycle, that's not a cause for concern.

Testing, ultimately, needs to be married to your experience in your body and with risk factors that you may have as an individual, like another diagnosis, family history, age, medication use, etc.

Testing should be able to help us rule something in or out, make a diagnosis, monitor progress or change over time, tell us something we don't know, screen for comorbidities and common concerns, and inform treatment. There will be added nuance here depending on who you are.

That all said, some parameters should be screened regularly in pretty much all menstruators. Routine check-ups like focused physical exams and testing are really important because they help to assess where your overall health is at, screen for common ailments, and track your progress over time. They also offer a great opportunity to address any questions or concerns you have about your health.

Routine testing can include your complete blood count, iron, glucose, cholesterol panel, cervical exams and cancer screening, kidney and liver function tests, blood pressure monitoring, and more. Then, there will be individualization depending on factors such as the presence of symptoms, age, medical history, family history, fertility goals, etc. that may incorporate STI testing, ultrasounds, mammograms, hormone levels, and such.

Condition-specific testing is discussed in the next seven chapters, but let's talk about a few tests that are relevant to just about all menstruators:

- **Ferritin and CBC**

  Ferritin is a test that is reflective of iron stores, whereas your complete blood count (CBC) tells us about your red blood cells, haemoglobin (oxygen-carrying protein in blood), and white blood cells.

  Iron deficiency is one of the most common nutrient deficiencies worldwide and can lead to anaemia, i.e., low red blood cell count or low haemoglobin levels. You need a consistent supply of iron to carry oxygen via haemoglobin in your red blood cells. This is especially relevant for menstruators losing blood through their periods regularly.

Symptoms like fatigue, sleepiness, shortness of breath, anxiety, hair loss, brittle nails, and low mood are commonly seen with low iron even if it's not flagged as being outside of the reference range. There was (finally!) an update to ferritin reference ranges in Ontario just in the last year to capture more people with low iron who were slipping through the cracks.

You can't make iron; it needs to be consumed through food or supplementation. Research, corroborated by clinical experience, tells us that a deficiency can exist irrespective of the kind of dietary regimen you follow (yup, you can be eating foods rich in iron and still be deficient).

If your iron is low, then diet alone might not be enough to bring up your levels and keep them there. Supplementation with one hundred to three hundred milligrams or more of iron for eight to twelve weeks and then retesting ferritin is the way to go. Ferritin levels should increase with regular supplementation in that timeframe.

Occasionally, iron supplements can cause constipation and other digestive disturbances; in this case, switching up the form of iron (perhaps to a polysaccharide, carbonyl, or bisglycinate form) may be helpful.

The one problem with relying on ferritin alone is that it can increase with inflammation in the body, giving a false impression that your iron stores are okay. This can mean that, sometimes, it's not the most accurate representation of iron stores, especially in an inflammatory condition like PCOS or endometriosis or even during an acute infection or inflammatory flare-up. In this case, your CBC, an iron panel, and inflammatory markers in serum can give you a better idea, especially if you have signs and symptoms of iron deficiency or anaemia.

- **Vitamin D**

  This is your sunshine vitamin that is formed thanks to a multi-step process requiring ultraviolet (UV) light to hit your skin for conversion.

  The maintenance dose in Canada is 2500 international units (IU) per day. It is assumed that Canadians, due to long winters and the latitude we reside in, should supplement with 1000 to 2500IU if symptomatic without even testing levels. If that sounds a bit like guesswork, then your assessment is accurate.

  Vitamin D has its hands in all of the body's different systems, from bone and immunity to hormone and mental health.

  Vitamin D deficiency can present as low mood, fatigue, musculoskeletal pain, and even general malaise or nebulous symptoms. It's associated with a wide range of chronic illnesses like infertility, menstrual conditions (like PMS, PMDD, endometriosis, and PCOS as you'll see in the following chapters), heart disease, osteoporosis, cancer, autoimmunity, and more.

  Many foods are fortified with vitamin D and it can be found in numerous multivitamin formulations. There are genetic differences in how people absorb and convert vitamin D. The best practice is testing instead of guessing so that your vitamin D levels are adequate, and also to ensure that you're not taking too much through supplementation since there is such a thing as too much vitamin D.

  In Ontario, testing used to be part of our provincial public healthcare plan called OHIP until the mid-2010s, at which point it was no longer covered for the general population. The exception is if you have osteopaenia, osteoporosis, kidney disease, and other extenuating circumstances like the use of medications that affect your bones. Testing for the rest of Ontarians costs approximately forty-five dollars and is not part of routine blood work. It can be requisitioned by your medical doctor, naturopathic doctor, or nurse practitioner.

- **Thyroid Function**

  Hypothyroidism (when the thyroid is under-functioning) and autoimmune thyroid disorders (when the immune system makes antibodies against the thyroid) are quite common in those assigned female at birth. Hyperthyroidism (overactive thyroid) is also possible.

  The screening test for thyroid function is called thyroid-stimulating hormone (TSH), which was discussed in Chapter 5.

  Even when patients are complaining of low thyroid symptoms like fatigue, hair loss, mood changes, weight gain, constipation, persistent bloating, menstrual changes, and feeling cold all the time, they are allowed to get worse before full testing or treatment is offered. Of course, we don't want to treat or medicate prematurely as TSH can naturally fluctuate for many different reasons like nutrition status, stress levels, body composition changes, illness, and pregnancy. Oftentimes, TSH will return to normal with none to little support, which is why repeat testing after an initial out-of-range test is standard practice.

  However, TSH also doesn't tell us everything either. It's a good starting point, but it isn't your actual thyroid hormones, T3 and T4. Moreover, it does not inform us of any antibodies against the thyroid. A big part of this lack of testing is the guidelines and algorithms that have been set forth for physicians to follow. The other part is that the treatment doesn't change conventionally whether you have antibodies or not.

  When people have clear signs and symptoms, more investigation and support should be offered. This is especially important when there is a family history of thyroid disease, unexplained infertility or decreased ovarian reserve markers, another autoimmune condition, and comorbid conditions such as PCOS. Autoimmunity of one organ is associated with other autoimmune conditions as well. As such, it is prudent for us to understand the full scope of someone's thyroid function.

Managing Your Hormone-based Concerns

Multifactorial conditions often require a multimodal approach that takes into consideration all of you. The next seven chapters contain considerations for some of the most common menstrual and hormonal health concerns in those of reproductive age, while the final few chapters of the book go through how you can improve your general menstrual and overall health.

I've picked some of the most prevalent hormonal topics, but not *all* of them. Each condition could have its own book(s)! You can skip to the diagnosis of interest or go through them all since many of us go through our menstrual life without an understanding of our health. You might just find that your experience is reflected in these pages. You can also learn about what treatment options are available to you, how you can start advocating to receive better healthcare, and how to start shifting the needle on your menstrual health.

A diagnosis can be helpful in many ways. It can be validating, provide language and direction, connect you with appropriate support and community, offer existing guidelines and treatment standards to help you with your health, and may facilitate referrals to other practitioners. On the flip side, it might put you in a box, limit the kind of care you receive, and have everything you are experiencing medically chalked up to your diagnosis; once you have a diagnosis, sometimes it can be quite challenging to get another.

All of these truths can coexist, including the fact that you are more than your diagnosis.

BOX 8. Why me?

As humans, we're curious. We're always looking for answers.

When it comes to health and disease, the questions often asked are: why me? how did this happen?

Reframed, these questions are often asking: what's the root cause? what did *I* do to cause this?

When it comes to communicable diseases like chicken pox or measles, we can pinpoint the culprit of your symptoms and treat accordingly.

When it comes to complex and chronic non-communicable conditions like endometriosis or PCOS, on the other hand, the answer is not straightforward as you'll see in the next few chapters.

There is a complex interplay between all aspects of your life, your environment, and even elements from before you were conceived. Then, there are dynamic and evolving factors over time.

This is why attempts to find the one underlying cause yield confusion and frustration. This is also why finding a singular definitive solution or "that one thing" often offers an incomplete strategy to all the presenting facets of a condition.

When looking at lifestyle medicine or how health conditions develop, we're often exploring relationships and patterns. That's why you'll see terms like "correlated to", "associated with", and "contribute to". We can often elucidate themes at the population level; however, depending on the study and subject, we can't hold the diverse, and sometimes contradictory, realities of an individual's health, life, and circumstance.

What this also means is that we often cannot speak in absolutes, especially when it comes to the complicated topic of hormone health. Layered atop that is the fact that science, humanity, health, and the world are constantly changing, which means there are shifts in how we understand and apply these interconnected concepts.

I know ... it would be easier if we could speak in absolutes! But, like life, health also operates primarily in the grey.

The way you eat didn't cause your PCOS or endometriosis, and it's not going to fix it either. Don't get me wrong: nutrition and gut health are important contributors to your PCOS or endometriosis, and nutrition is an extremely useful tool in our management toolkit. However, it's one facet of many.

We're still learning and understanding the nature and interplay inherent to these kinds of medical conditions. Focusing on the root cause here can feel like a losing battle. It can even delay effective strategies to help you feel better and move the needle on your health so you can live your life.

Just like health and prevention are multifaceted and require us to engage in a variety of health habits, disease management is similar.

You'll likely go through seasons of ebb and flow in how you experience and relate to your condition. Your definitions of health and wellness will evolve. Furthermore, you'll have to consider what the cost of doing nothing is and when you are "healthy enough". There's a time to ramp things up and a time to maintain or coast.

In the next chapters, I've laid out conventional and integrative options as well as lifestyle practices such as nutrition and exercise–these therapies do not form an exhaustive list and many of them are not mutually exclusive.

The pendulum in the medical, as well as wellness industries, can swing wildly at times, from one extreme to the other, often propelled by oversimplification, misinformation, and the polarized "us versus them" rhetoric commonplace on social media. There is nothing wrong with choosing a strategy that works for you and fits your life, whether that is allopathic, naturopathic, or a combination. Moreover, just because something is categorized under conventional medicine, does not mean it's also not naturopathic, and vice versa.

I've focused mostly on therapies that have some degree of scientific research available and ones that are most prevalent in North America. Many treatments have not yet been studied well or at all, there might be treatment options that will become outdated in the future, there are therapies that have only been studied in animal models or *in vitro* thus far, and there are also traditional and cultural remedies, and energetic and spiritual forms of healing that I know little to none about.

The body sometimes works in mysterious ways. It can manifest symptoms in a manner that confounds modern medicine, but those familiar with the energetics and somatic expressions of emotion and trauma—our own but also of our ancestors—have a deeper knowing of the interconnectedness of everything.

Scientific research tends to be quite reductive; the scientific process requires that. By breaking things down systematically, we gain clarity. We've learnt so much about hormones, medical conditions, drugs, herbs, food, antioxidants, vitamins, and other therapies in this way. These studies help us elucidate the scenarios in which a particular treatment will be most beneficial or super risky. They also help us discover a wider range of effective treatment options while broadening our understanding of the body as well as the similarities and differences between us.

That being said, science-backed strategies and traditional wisdom are not mutually exclusive, even though they're often presented as such. Traditional and cultural medicines and practices, even as they evolve, tend to be more wholesome and less extractive than what science and modern medicine require of therapies. Just because something doesn't have a double-blind randomized control trial (yet) does not mean we should disregard it. Unfortunately, safe and effective therapies are dismissed and even labelled as quackery for this reason quite frequently. Medical research is covered more comprehensively in Chapter 23.

Remember, there is a place for your cultural and traditional customs in your health and healthcare; they simply may be out of the scope of this book as well as that of your healthcare providers.

There is also space for exploration and curiosity. There is room for you to feel a wide spectrum of emotions and to adapt as you experience life and the world in real time. And, there should always be a place for your values, life goals, financial situation, respect, and dignity in your treatment plans and doctor's office.

Shared decision-making is how your care should be planned. This requires conversation and time, which are limited in our medical care system. Having time to discuss your health during a medical appointment is the evidence-based way to offer healthcare, especially for complex chronic conditions and comorbidities. These situations often require some trial and error as well as creativity in clinical care.

When a treatment option for a certain condition is discovered that works for a large percentage of people, there *will* be exceptions. What works for someone might not work for someone else. And, that's okay, because there are multiple ways to get to better health. "(Better) Health" is a moving target with an ever-changing definition anyway.

I used to have a lot of apprehension and anxiety around all of this. Now, this often forms an entry point to honest and open discussions that not only facilitate clinical decision-making but also foster curiosity, rapport, and agency.

My practice is ever-evolving as I further my knowledge, grow as a clinician and person, make mistakes (it's inevitable!), and work with real people. It's cliché to say, but I'll say it anyway: the more I know, the more I realize just how much I don't know. What I do know is that many things can be true at once, that health and life are not linear journeys, and remaining curious is critical to growth. I've ditched some dogma and become more flexible in my clinical care (and life!) over time.

I reflect regularly on how to best honour the person in front of me and the practicalities of their life, combining the art and science of medicine. Naturopathic medicine combines conventional medical practices as well as holistic and traditional medicines. My profession has devalued, appropriated, and misrepresented knowledge and practices from around the world–as has conventional medicine. I can't change that past, but I can commit to addressing this part of medical history as I continue to work, learning, unlearning, honouring, and advocating for reconciliation and social justice.

Naturopathic medicine is a regulated profession where I live in Ontario as well as British Columbia, Alberta, Saskatchewan, Manitoba, and the Northwest Territories (and other places, like some states in the US). What this means is that the profession has rules and regulations in place with a governing body for public protection just like any other healthcare profession including medicine, physiotherapy, nursing, chiropractic, etc. The designation of naturopath and naturopathic doctor are protected titles in these jurisdictions. You need to have successfully completed education at an accredited school followed by standardized national and provincial and state exams, while keeping your license and professional requirements up-to-date. In unregulated areas (and in the online world), there could be people using these titles falsely or inappropriately. Most local or national professional regulatory bodies and professional organizations will have a roster of practitioners with an active license that you can consult with.

This is as good a place as any to remind you that I'm not your naturopathic doctor (unless you're a patient of mine, in which case I am your ND but this book does not constitute individualized care or a doctor-patient relationship!).

There are myriad treatment options available for people with hormone-based conditions. Some treatment considerations listed in this book might not be appropriate for you. Also, the existence of numerous options doesn't mean that you employ *all* of them. The selection is based on your unique whole self. There is room for conventional, traditional, and integrative therapies; since there can be interference at times, it's best to ensure the compatibility of your plan with your healthcare team.

We oftentimes lack the scientific evidence employing a few different therapies together, or how we approach your care if you have three different diagnoses that we're trying to manage. This is where medical decision-making can get murky. Ultimately, nuance and individualization are needed. Sometimes, a bit of trial and error is required. There is a science and art to medicine.

We need to make sure we're reevaluating or measuring your progress appropriately, pivoting as needed. We need to make sure that everyone's on the same page as well: you might be expecting 100 percent resolution of your symptoms like no period pain when you have fibroids, while 40 to 70 percent improvement is what we're expecting medically with the chosen treatment.

We also have to keep in mind that sometimes, treating one aspect of a multifaceted condition can lead to new or unexpected outcomes in other areas. For example, shrinking a precancerous growth over three months is great; however, it can come with debilitating fatigue, severe digestive upset, and poorer quality of life. Taking the treatment comes with consequences, as does not taking the treatment. Some patients have to choose between not-great options only.

In an ideal world, you'd be given a buffet of options for shared decision-making with your health team, starting somewhere and reassessing as you go. If you're feeling unsupported or unvalidated in your medical care, there are tips and tools for self-advocacy in Chapter 31.

The way you view health, conduct risk-benefit assessments for yourself, and make health decisions can and *should* change over time. In the same vein: as we build upon our current knowledge of the menstrual cycle and hormones, your healthcare team and the care you receive should reflect these changes as well (hopefully with more diversified options that include community care, natural and holistic treatment modalities alongside pharmaceuticals, and systemic change with decolonization and infrastructural revamping).

CHAPTER HIGHLIGHTS

- Menstrual health is multifactorial. Factors like lifestyle, nutrition, mindset, genetics, upbringing, healthcare, the environment, luck, and exposures while you were in the womb all play roles. What this means is that there are some things under your control and others that are not.
- Even if we all did the same things in the same way at the same time, we'd all have different experiences through all of that and end up in different places.
- Hormone testing doesn't always give us the answers we are looking for. Testing must always be related to you, your body, and your experience.

- Routine testing can be helpful for screening, measuring progress over time, identifying contributing factors to any concerns, and informing your treatment plan. For menstruators, blood parameters such as iron, vitamin D, and thyroid function should be tested routinely as they are related to many menstrual health conditions.
- Conventional and more natural and integrative therapies can be used together in a shared decision-making process. It's best to confirm the compatibility of these strategies with your healthcare team to minimize interference and/or harm.
- There should always be space for your background, life experience, current situation, values, culture and tradition, curiosity, and respect in your healthcare.

CHAPTER 15
Painful Periods

Period pain is one of the most common menstrual concerns globally. In medical speak, pain and cramping associated with menstruation is called dysmenorrhea.

There are two categories of dysmenorrhea:

- Primary dysmenorrhea, which is when menstrual pain is not caused by an underlying organic disease. This is the focus of this chapter.
- Secondary dysmenorrhea, which is associated with an underlying issue like endometriosis and fibroids (covered in Chapters 20 and 21, respectively).

Dysmenorrhea can feel like throbbing, cramping, and pain in the lower abdomen right before and/or during your period. It can radiate to the thighs and lower back. This pain can last anywhere from a few hours to a few days, typically worsening on the first or second day of your period when the flow is heaviest. There is a clear and predictable pattern of pain in dysmenorrhea: it starts right before or during your period and resolves as your flow starts to peter out. If the pain persists for longer, or is present at multiple times during the menstrual cycle, it might point us in the direction of secondary dysmenorrhea.

It can be accompanied by heavy bleeding, significant clotting, lower back pain, nausea, vomiting, diarrhea, leg pain, headaches, migraines, fatigue, and general malaise. It's classified as mild, moderate, and severe. It can impact sleep, mood, ability to function, and overall quality of life. Moreover, it's correlated with premenstrual syndrome (PMS), meaning that menstruators who experience PMS have higher rates of dysmenorrhea and vice versa.

Primary dysmenorrhea usually starts in adolescence, typically within two years of menarche, when the chances of structural changes and other conditions are a bit lower. Some people have it right from the get-go, while others develop it shortly after beginning their menstruating life.

We don't fully understand why some people don't get any pain whatsoever, while others have debilitating pain, but we have elucidated some correlations and contributing factors.

During a period, endometrial shedding happens as the consequence of some pretty fascinating physiology which was covered in Chapter 3. It involves a drop in your sex hormones and the creation of an inflammatory response. This inflammatory response involves increased blood flow to the endometrium, higher fragility of blood vessels, and some clotting processes. It's mediated by compounds called prostaglandins and leukotrienes.

The leading theory in primary dysmenorrhea is that there is an overproduction of prostaglandins. This has been determined in research studies that have quantified inflammatory compounds in menstrual flow.

Increased prostaglandins lead to a more pronounced inflammatory response as well as more contractions in the muscular layer of the uterus called the myometrium. That's why painkillers like ibuprofen help with period pain: they inhibit the production of prostaglandins.

Other risk factors for dysmenorrhea include family history, heavy menstrual flow, earlier age at menarche, nutritional deficiencies, gastrointestinal disorders like irritable bowel syndrome (IBS), stress, and trauma.

Pain chemistry changes may also be implicated in primary dysmenorrhea. Essentially, there might be an increased sensitivity to pain signals over time.

Remember: some discomfort or feelings of pressure during your period are considered fairly normal–a lot is going on. However, pain for which you need to take multiple extra-strength painkillers, use a bunch of different management strategies, or miss out on work, school, and life events, should be addressed with a healthcare provider.

Primary Dysmenorrhea Assessment and Diagnosis

Unfortunately, period pain has been normalized even when the pain is agonizing. This gets in the way of menstruators accessing appropriate care.

As with any health concern, a comprehensive evaluation of your experience, psychosocial factors, past medical history, blood work, physical examination, and imaging will allow for the most accurate diagnosis as well as care that fits your needs.

There is no specific blood test that is diagnostic for primary dysmenorrhea. The goal with blood tests is to see if there are aspects of your health, especially those related to increased period pain, that we can address. Vitamin D deficiency is associated with period pain, so testing for it and ensuring you have enough is important. If your period pain is accompanied by heavy bleeding, then ruling other conditions in or out to discern primary vs. secondary dysmenorrhea is also prudent. Imaging like ultrasound technology can aid in this in many cases. We would expect imaging to be unremarkable (meaning that no pathology is detected) in primary dysmenorrhea.

An evaluation of your nutrition, sleep, stress, and activity levels is often missed conventionally. These salient factors impact dysmenorrhea as well as pain and inflammation in general as you'll see later in this chapter.

What you can do–and only you can do this–is take a full inventory of your pain. The questions to consider here include:

- When does it start and end?
- When was the first occurrence ever?
- What kind of pain is it (sharp, shooting, pressure, cramping, etc.)?
- Where is it located?

- How intense is it (rate it on a scale from zero to ten)?
- What are the accompanying symptoms?
- What makes it worse?
- What helps?
- What have you tried that doesn't help?
- How does it impact your life?
- How has it changed over time? If it has changed recently, what's different in your life and health?

These may seem like a lot of questions; however, once you have the data, it can be conveyed to your doctors succinctly in a few sentences.

If you use painkillers, tell your doctor the amount and strength you're using. Sometimes we forget that menstrual pain is bearable because of painkiller use. I've had patients in the past who take heroic doses of painkillers to manage their period pain–this deserves to be addressed more comprehensively! Another consideration is to take note of any digestive issues as gut problems like loose stools, cramping, and gas can be linked to period pain especially if you have a condition like IBS.

A referral to a gynaecologist, especially if your family doctor isn't offering you the support you need and deserve, is something to advocate for. Sometimes, secondary dysmenorrhea exists in adolescents but is not captured due to a lack of investigation in this population. This is a common occurrence for patients who later find out they have endometriosis.

Since dysmenorrhea has been historically minimized, the onus (whether we like it or not) falls on us to advocate for ourselves if we want and need support.

Pharmaceutical and Conventional Treatment Options

The most common treatment for dysmenorrhea that you are likely familiar with is NSAIDs (non-steroidal anti-inflammatory drugs). You know these as painkillers like ibuprofen, naproxen, and mefenamic acid. These meds help by decreasing those inflammatory prostaglandins mentioned earlier.

They're quite effective, especially when used one to two days leading up to your period. They can also help with symptoms associated with menstrual pain like heavy flow and headaches.

The downside of using NSAIDs, especially when taken frequently, long-term, and at high doses, is irritation and damage to the gut, kidneys, and liver.

Other pain-relievers like acetaminophen, sometimes combined with anti-histamines, may also be helpful. Histamine, the immune compound associated with allergies and inflammation, is also involved in the changing endometrium and inflammatory response of menstruation. It may be associated with the general malaise some menstruators feel at the end of the luteal phase and the beginning of a period.

Oral contraceptives and the hormonal IUS are other commonly used treatments for period pain. They have the added benefit of providing contraception and, similar to NSAIDs, they do tend to help with heavy flow and other hormonal symptoms as well. We're adding hormones to the mix, so there can be a variety of unintended effects that are discussed in Part Three of this book; however, they can provide blanket coverage for you if that's the path you'd like to take.

Other treatment options include ablation therapy, embolization therapy, and surgery; surgery is more common for secondary dysmenorrhea due to endometriosis and fibroids.

Nutrition and Dietary Strategies

Many lifestyle medicine strategies can be implemented as part of a primary dysmenorrhea treatment plan. Moreover, these elements also improve overall health while reducing the risk of chronic illnesses. These include staying hydrated, limiting alcohol intake, and having diverse sources of nutrition.

A higher degree of period pain is associated with caloric restriction as well as skipping meals. When you eat less as a whole, you're also going to get fewer nutrients like antioxidants, fibre, vitamins, and minerals. In fact, there is an association between lower period pain and a higher consumption of fruits, vegetables, dairy, and essential fatty acids.

Low intake of omega-3 fatty acids is associated with many pain disorders. Biochemically, these fats are anti-inflammatory with antioxidant properties, helping resolve inflammatory processes in the body. They also help with the fluidity of our cell membranes, which is important for how our cells function (think back to grade school science!).

In the same vein, there is an association between a higher degree of period pain and low omega-3 intake. The flip side is also true: high dietary omega-3 intake is associated with lesser pain. Studies have shown that omega-3 consumption can help decrease the rescue dose of ibuprofen needed to reduce dysmenorrhea.

The dose of omega-3s is a tad difficult to discern since there is heterogeneity in the studies regarding duration, dose, and baseline dietary omega-3 intake. Supplementation seems to be the most helpful when your baseline intake is low. What this likely means is that if you have no or low dietary omega-3s in your day-to-day, you probably will benefit the most from adding more omega-3s to your life.

Essentially, the key to less painful periods involves eating lots of plant foods, focusing on a balanced diet with nutrient-dense foods, making sure you're getting some omega-3s, and eating enough for your body and its energy requirements.

Exercise and Physical Modalities

As you work on diet and nutrition, you can layer in other habits like movement and exercise, both of which can help with dysmenorrhea.

Anything is better than nothing when it comes to exercise. Formal exercise, whether that's cardio or resistance training, has benefits beyond dysmenorrhea (discussed in Chapter 30).

My general recommendation is usually to do whatever you are able to do, as any form of movement–from walking and dancing to weight-lifting and biking–will be helpful. Stretching, yoga, and isometric exercises, especially ones that involve stretching the back and bending forward, can be of benefit here.

Even passive mobilization of the pelvic area through chiropractic care, massage therapy, and osteopathy is beneficial. There are two specific health tools that I want to highlight:

- The first is good ole heat; warm baths and heating pads are used widely for menstrual cramps. I'll offer a word of caution with heat: I've seen patients, typically undiagnosed endometriosis ones, with heat burns on their skin from using a heating pad too often or for too long. Please be careful. As mentioned earlier, it is possible to do harm even with treatments that are seemingly safe and simple.
- The second tool is TENS, which stands for transcutaneous electrical nerve stimulation. This therapy utilizes electrical impulses that are relayed across your skin via sticky pads placed on your lower abdomen and/or lower back for dysmenorrhea. A physiotherapist should be able to help you with this to determine pad placement, duration, frequency, and intensity.

Another kind of physical therapy called pelvic physiotherapy can be of benefit here as well. As discussed in Chapter 1, the pelvic floor is the group of muscles to which the reproductive organs are anchored. If these muscles are tight, lax, and/or uncoordinated, this can contribute to pain. A pelvic physiotherapist typically conducts an external and internal exam to determine the health of the pelvic floor and offer treatment based on their assessment.

Last, in this section, but not least is acupuncture, which is helpful not only for period pain but also for other menstrual and health issues. Traditional acupuncture treatments rely on a thorough assessment of the whole person, a diagnosis, and a selection of points for needling based on that. This age-old therapy can have a positive impact on the quality of life and symptoms of those with dysmenorrhea, both primary and secondary.

Lifestyle and Stress Management

Your period health plan should always incorporate stress management, rest, and getting good-quality sleep.

Unmanaged stress and poor sleep can exacerbate period pain and inflammation. Pain tolerance can decrease when we lack good-quality sleep. Energy levels, mood, the capacity to take care of yourself, and metabolic regulation, i.e., your hormonal rhythm, can also be affected.

So, don't forget the importance of those in your health. Recruiting extra help, delegating or delaying tasks, sharing your experience with those you live with, and building in more recovery time can go a long way, as can adding supplements and herbs, which we'll get into next.

Herbs and Supplements

First up: is good ole ginger. Ginger is an extremely effective herb for dysmenorrhea. It can be taken in a variety of different forms: powder, capsule, tea, and fresh root.

Multiple research studies have shown oral ginger supplementation to be very effective for dysmenorrhea–comparable to NSAIDs like ibuprofen and mefenamic acid; and that's without the ill digestive effects of NSAIDs. Ginger is an anti-emetic, meaning it can help with nausea and vomiting. It can cause a bit of heartburn in some individuals, but other than that, it's tolerated fairly well.

Ginger works by decreasing the production of the inflammatory compounds called prostaglandins and leukotrienes and has additional constituents that exert antioxidant and analgesic, i.e., pain-relieving, effects. As a cherry on top, it can help with heavy bleeding too.

It helps with the severity of pain even if started at the onset of the period, but it seems to only be effective at decreasing the duration of pain if you start before the onset of menses, similar to NSAIDs.

The dosages studied include 500 milligrams of encapsulated ginger powder three times a day or 250 milligrams four times a day. So, we're looking at a total dosage of 1250-1500 milligrams in divided doses of standardized extract twice daily starting two to three days before the onset of the period until the end of the heaviest flow (which is probably day two or three for most of you).

The nice thing about this herb is that instead of taking it all the time, it can be taken cyclically. Ginger is a big winner for period pain in my books!

BOX 9. Heat, heritage, and herbalism

From a more traditional and herbal medicine perspective, ginger is known as a warm herb. My mum and relatives always used to refer to foods as "garam" (warm) and "thanda" (cool).

I didn't fully understand these concepts until I got a bit older, especially when I realized that these energetics of food are common in many parts of the world. (Plus, I was too cool to listen to my mum!).

In Traditional Chinese Medicine (TCM), there are warm, neutral, and cool foods as well.

The diagnostic pattern associated most with dysmenorrhea is stagnation, which is when Blood and/or Qi are not flowing smoothly. (Qi, loosely, is your vital energy or life force).

The therapies listed here–heat, movement, food, ginger, acupuncture, rest–all fall into the nourishing, moving, and warming category to promote the smooth flow of Blood and Qi. Since cold or cooling foods and practices can impede flow, these are to be minimized. This includes icy beverages, cold showers, and consuming a lot of raw foods.

This is a solid example of how scientific research and age-old medicine converge!

Other herbs that can be of use for dysmenorrhea are cinnamon, *Boswellia serrata* (Indian frankincense), cinnamon, chamomile, and crampbark.

Cinnamon can inhibit prostaglandin biosynthesis. Similar to ginger, it is also a warming herb. Standardized extracts have been shown to decrease the severity and duration of pain but not as effectively as ibuprofen.

Boswellia serrata is a potent anti-inflammatory herb that has historically been used for joint pain. It can help decrease dysmenorrhea especially when accompanied by heavy menstrual bleeding. The scientific data on this herb isn't very strong in quality or quantity, but it is often used in capsule or liquid form.

Chamomile is an herb that many people are familiar with for its soothing effect on the nervous and digestive systems. It contains anti-inflammatory and anti-spasmodic (i.e., anti-cramping) compounds. Even though the effect on dysmenorrhea is a mild one, it can be added in as a gentle support. Unless there's an allergy to chamomile or related herbs, it's fairly safe. Two cups per day of chamomile tea may help with both, cramping and PMS.

Crampbark, similarly, offers its support by working on inflammation and–as the name suggests–cramping. It has been used in herbal medicine for a long time, typically in combination with other herbs (including the ones covered here as a tea or tincture, typically).

Topical creams and oils with medications, cannabinoids, herbal extracts, and essential oils may have utility as well. Scientific literature is sparse in this area, however, we borrow a bit from other pain conditions. We also take from traditional and herbal medicine practices that have utilized oil infusions, poultices, and balms along with heat to support menstruators with abdominal and back pain. As always, we want to ensure the safe and judicious use of topicals, especially if there are any essential oils involved. Unsafe and undiluted use of essential oils has increased dramatically in the last decade.

Moving on from the herbs, there are minerals like magnesium. Magnesium is a pretty popular supplement widely used in the health and wellness space. Dosed at approximately 200 milligrams, magnesium may help by relaxing smooth muscle. It's also involved in hundreds of biochemical reactions in the body. It has been shown to decrease the amount of painkillers needed for period pain, and can also help with menstrual migraines and mental health symptoms that might come up during this time.

Other interventions that have limited evidence behind them as of 2023 include supplementation of nutrients like vitamin E, zinc, calcium, and B vitamins. The dosages studied are slightly varied, the studies are short-term, and some didn't account for dietary intake, so it's unclear what baseline levels were like in the populations researched. These studies are primarily in adolescents. What we know about adolescent nutrient intake is that it could be more robust–there's often food restriction, picky eating, changing bodies, skipping meals, and irregular eating patterns.

My takeaway is that we need to do a better job at getting nutrients, aka food, into young people who are the most prone to primary dysmenorrhea. A multivitamin or multimineral can be considered but should be discussed with a practitioner who can offer clear guidance.

CHAPTER HIGHLIGHTS

- Dysmenorrhea is the medical term for pain associated with periods. It can feel like cramping, throbbing, and/or aching in the lower abdomen that can radiate to the legs and lower back. It is cyclical, starting close to the beginning of a period, typically worsening on the heaviest day of flow.
- Primary dysmenorrhea is when period pain is not caused by an underlying issue.
- Secondary dysmenorrhea is when period pain is a result of another condition like fibroids or polyps.
- We don't fully understand primary dysmenorrhea, but the leading theory is that there is an overproduction of inflammatory immune compounds called prostaglandins. Imaging is typically unremarkable and there are no diagnostic tests. What's most important is taking inventory of your pain and lifestyle and relaying that to your doctor(s).

- Conventional treatment options include NSAID painkillers like ibuprofen and naproxen that decrease prostaglandin production. The birth control pill and hormonal IUS are also used often, especially if contraception is also required.
- There is a correlation with vitamin D deficiency as well as lack of sleep, high stress, and lower activity levels.
- Physical modalities like massage, chiropractic, and physiotherapy (including TENS and pelvic physiotherapy) have been shown to improve period pain. Acupuncture can also be of benefit.
- Ginger is the herb with the most amount of research behind it for dysmenorrhea as well as backing from cultures and traditions across the world. Other beneficial herbs include Boswellia, cinnamon, and crampbark.
- Multiple vitamin and mineral deficiencies as well as poor and irregular dietary habits are implicated in primary dysmenorrhea. Consequently, working on nutrition and perhaps supplementing with a multivitamin may be helpful.

CHAPTER 16
PMS

PMS: these three letters are thrown around fairly, especially as a pejorative to menstruators. While this connotation pretty much sucks, PMS, or premenstrual syndrome, is very real.

PMS occurs in the luteal phase of the menstrual cycle. The symptoms are usually physical, somatic, and/or psychological. It can range in severity, and the number of symptoms for this diagnosis is unspecified.

Signs and symptoms can include bloating, acne, headaches, cravings, fluid retention, severe pain, breast tenderness, mood swings, numbness and tingling down the legs, migraines, headaches, nausea, debilitating fatigue, irritability, insomnia, and more.

Premenstrual symptoms arise due to hormone change. Not everyone responds to these shifts in the same way, though.

How your body perceives hormonal fluctuations is a crucial, and slightly nebulous, piece of the PMS puzzle.

Where we currently stand is that your hormones are happening to you, while you and your life are happening to your hormones. This is especially true at hormonally-sensitive times like puberty, pregnancy, postpartum, and perimenopause, and when there are high amounts of prolonged physiological change and stress. What this means is that your experiences and everything else happening in your life and body impact your hormones and symptoms, which, in turn, informs how you experience your hormones. Then, we have to layer in some preliminary research indicating that estrogen in the follicular phase affects us a bit differently in the luteal phase (which is fascinating!).

When the stress system, the HPA (hypothalamic-pituitary-adrenal) axis, is activated significantly, your body "marries together hormone change and stress", as my colleague Dr. Jordan Robertson, ND, put this in a webinar once.[7] As a result, a subsequent hormone change can trigger a stress response.

We know that early childhood abuse, high stress in adolescence (especially during puberty), trauma around the time of our conception, and negative associations with menarche are all risk factors for future PMS. It makes sense that colleagues and I saw an uptick in moderate to severe PMS as the world navigated life during the COVID-19 pandemic, power politics, and geopolitical strife.

As always, there are other contributors. As we age, we can become more sensitive to our hormone changes. This makes sense for many reasons like cellular ageing, more lived experience in our bodies, changes in hormone production, cumulative effects of stress, etc.

Lifestyle factors are mentioned later in this chapter, but I want to bring up a major theme in those with severe PMS: when we overdo it in the first half of the cycle, we see a more symptomatic second half of the cycle.

When we're in go-go-go mode all the time, premenstrual and menstrual concerns seem to get worse. It's a pattern I see in myself and it's a pattern I see in my patients.

I get it, if you tend to feel unwell in the luteal phase, then you might put extra things on your to-do list in the follicular phase. But this tends to become a vicious cycle: overdo it, crash, overdo it, crash. We can't keep up an unsustainable pace in life indefinitely. One phase of the cycle flows into the next.

---

[7] *Robertson, PMS & PMDD Webinar, 2021.*

PMS Assessment and Diagnosis

A diagnosis of PMS is given when there are:

- at least three days premenstrually when family life, social life, and the ability to work are impaired, *or*
- family life and social life is impaired for five days
- symptoms are present in at least two cycles during the luteal phase, but they resolve within the first couple of days of your period, no longer being an issue in the follicular phase

Technically, PMS is a diagnosis related to ovulation, i.e., you can only have PMS if you've ovulated. So, if you're on the combo oral contraceptive pill, have had a bilateral oophorectomy and have no ovaries as a result, or are on hormones that are suppressing ovulation, then it's not *technically* PMS. This doesn't mean that you can't have premenstrual or PMS-like symptoms because you totally can—we see this in those of you who are on the pill, in perimenopause, during anovulatory cycles, and in trans women who are on hormones. Fluctuating hormones can manifest in a variety of ways.

When we pool worldwide evidence, the prevalence of PMS in menstruators of reproductive age is approximately 48 percent.

PMS can last anywhere from four to fourteen days, significantly impacting daily life. This means we have *a lot* of menstruators who aren't showing up for themselves, their lives, and their loved ones for a significant chunk of time.

We need better public health and education measures in place. We have pilot programs across the globe where we see an improvement in menstrual symptoms simply when we have better education and awareness in place through community and school programs. The same would happen if we had more doctors than we currently do who *actually* believe patients when they say that their symptoms are getting in the way of their lives, especially considering the evidence linking better health outcomes and patients feeling heard and validated.

This point is a huge one because, historically, and even now, the experience and quality of life of the menstruator don't hold as much value as they should. There are many research studies from the last two decades that have surveyed physicians about the importance of subjective data obtained by patients; over 50 percent of practitioners indicated that it was not very important. Sure, some of this research is a bit dated, but we have so much global anecdotal data just from this last decade that corroborates this narrative, especially concerning stigmatized and marginalized folk.

Then we layer in the question of: Whose job is it to treat PMS anyway? PMS symptoms span an array of bodily systems and medical specialties. So, who do we turn to? Family doctors are often ill-equipped to help us with the various facets of premenstrual conditions, especially if they're not listening to us or believing us. They're also the gatekeepers of specialist referrals in Canada. Doctors listening and remaining curious can also help tease out whether a certain set of symptoms is PMS or something more severe, like PMDD (covered in Chapter 17).

A PMS workup should include a thorough history, including the nature of symptoms, severity, and timing, and how these symptoms impact your life. This is where your tracking data from Chapter 7 comes in!

For those of you wondering about hormone testing: no blood test is going to diagnose PMS. Ten people could have exactly the same hormone levels and have vastly different premenstrual experiences. Conversely, another ten people could have exactly the same PMS symptoms but have varying hormone levels upon testing.

This doesn't mean that blood work has no role here. It helps us rule in contributing factors like vitamin D deficiency, and to rule out other issues like a thyroid problem. Relevant blood work for PMS can include CBC (complete blood count), ferritin (stored iron), vitamin B12, vitamin D, thyroid panel, prolactin, and other parameters relevant to you.

A digestive work-up for any of you with significant gut symptoms and bowel disturbances is appropriate as there is a relationship between IBS (irritable bowel syndrome) and PMS. Oftentimes, the luteal phase is the trigger for initiating a more thorough investigation into other diagnoses.

Pharmaceutical and Conventional Treatment Options

We need better medical education in general as well as better continuing education at the level of the family doctor on PMS. The latest data show that non-pharmacological interventions should be the first line of treatment; counselling, nutrition, exercise, social support, adequate rest, heat applications like hot baths, vitamins, supplements, and cognitive behavioural therapy (CBT) should be at the forefront of PMS care.

Unfortunately, these aren't family doctors' areas of expertise, especially in Canada. Consequently, many people end up with birth control pills and antidepressants as first-line treatments without discussing things like validation, energy availability, nutrition, stress, movement, and social support.

Lifestyle and Stress Management

Working on your mindset and mental landscape can be extremely helpful through counselling, connecting with others, self-care, and creating space to simply be. Shifting language to be more positive or neutral, taking things off your plate (figuratively, not literally!), eating enough, building in extra time for yourself, focusing on good quality sleep, and talking about your experiences can all be beneficial.

In fact, in a study done in couples with one menstruating partner, it was shown that when we share our experience with our partners, PMS gets better. This is true for me, and I know this is true for many of my patients. It also makes sense: sharing our needs, feeling heard, and asking for help are all parts of healthy relationships.

Oftentimes, we need to spell out what that means so the people in our lives can support us the way we want to be supported, or to simply be understood. I know it might not always work exactly the way we want (I can speak from experience!), but it *is* still worth it.

Even sharing negative premenstrual and menstrual experiences, and bonding over them with other people, has been shown to lessen some of the mental-emotional symptoms of PMS. This can even be online through blogs, forums, and other communities (beware of online trolls, though!).

Connection is vital to health and vital to life.

I love connecting with nature, especially in the late luteal phase when my energy starts to dwindle a bit. I find that this is super supportive of my mental and physical well-being. It also allows me to slow down, connect with myself, and gain some perspective on life.

Spending time in nature, called forest bathing from the Japanese practice of *shinrin yoku*, helps reduce the stress hormone cortisol, uplifts the mood, reduces anxiety and irritability, and decreases pain perception. These are some of the issues experienced in PMS, so I borrow from that data and apply it to this phase of my cycle and encourage patients to do the same.

Other practices and therapies that help with stress management and relaxation like massage, exercise, meditation, baths, hobbies, and spending time with loved ones (including pets!) can also help with PMS. I love cuddling with my two furry friends any time I'm down. Acupuncture and any form of movement and exercise are also effective strategies for all aspects of PMS.

Nutrition and Dietary Strategies

Under-fuelling or restricting calories in the luteal phase can worsen PMS. If you're not eating enough for your body to feel and function well in the first place, then you may just end up with more anxiety, fatigue, irritability, headaches, and perhaps more night-time cravings since you simply didn't eat enough during the day. Hormone change in the luteal phase can be the fuel to stoke that fire.

Moreover, research indicates that resting metabolism might (very mildly) rev up in the second half of the cycle. This means that you may need anywhere from twenty to over two hundred calories per day at this time depending on activity levels, body composition, and more.

The bottom line here seems to be that you need to eat enough (and eat all your macronutrients including carbs and fats) to feel well and have less PMS. This way you'll also get most of your vitamins and minerals through your diet, which is a requirement for peaceful periods. You'll also feel satiated instead of ravenous if you make sure you're eating regularly throughout the day. For more on nutrition, check out Chapter 29.

The types of foods that can worsen PMS symptoms, especially bloating, breast tenderness, and inflammatory symptoms, include trans fats, high amounts of saturated fats (typically found in animal products and coconut oil), refined sugars, and certain processed foods that are not particularly nutritious. There are likely a few considerations here: these foods are typically lacking nutrients, hyper-palatable meaning you might overconsume them, can lead to an increased inflammatory load on the body, contribute to dysregulated blood sugar levels, might be indicative of lifestyle and socioeconomic contributors, and maybe replacing sources of fuel that offer more macro- and micronutrients that would support a smoother PMS time.

Foods beneficial for PMS include fibre- and antioxidant-rich foods like blueberries, watermelon, pomegranate, and blackberries.

The other specific plant foods we have research for are ground flax seeds (approximately fifteen to thirty grams, i.e., one to two tablespoons) and soy protein (approximately ten grams, i.e., one cup of soy milk or half a cup of tofu) to improve breast tenderness, bloating, and hot flashes premenstrually.

Last, but certainly not least, routine dietary omega-3 consumption a few times per week can help with the physical and psychological symptoms of PMS. Supplementation might be beneficial for those of you who consume little to none in your diet, but food can offer sufficient omega-3 fatty acids for most of us.

Alcohol worsens PMS, especially anxiety, insomnia, headaches, and breast tenderness. This holds when it's consumed at any part of the cycle, but especially when you drink it in the luteal phase. Eliminating, or at least decreasing, your alcohol intake can be a good measure to put in place. If you're someone who turns to alcohol to unwind during your PMS, replacing it with something else to help you unwind can be an important consideration.

Similar to alcohol, caffeine is something that many menstruators turn to during the premenstrual time, especially if fatigue is weighing you down and you've got things to do. However, caffeine consumption of two or more cups per day is associated with increased PMS symptoms like bloating, anxiety, and breast tenderness. You don't have to get rid of all your coffee, but decreasing it is helpful. Try to stick to one cup in the morning–perhaps one and a half–and replace your second (and third and fourth!) cups of the day with a decaf; or better yet, opt for a green or herbal tea like chamomile, which we'll get to shortly.

Herbs and Supplements

We'll kick off with our favourite, which is good ole vitamin D. Research shows that low levels are associated with PMS. When vitamin D levels rise to adequacy, PMS gets better. Testing and appropriately supplementing is simple and helpful for the cognitive function, anxiety, fatigue, and cramping of PMS.

Next up, we have calcium, which fluctuates throughout our menstrual cycle in relation to our hormones. Increasing dietary calcium (especially if low) and/or supplementing calcium beyond dietary intake can both help with mental-emotional PMS symptoms as well as some physical ones. What this means is that if we don't get enough calcium in your diet, increasing calcium intake helps; and if we are getting enough calcium, additional supplementation can also help. The dosage in studies is about 500 to 1500 milligrams per day, and it can be used cyclically, i.e., just in the second half of the cycle.

There is some data on magnesium, which can help with symptoms like migraines, bloating, pain, and breast tenderness. It also helps with subjective measures of general anxiety in people with PMS, so it might be a consideration here. The dosage for magnesium lies between 100 and 400 milligrams per day, usually in divided doses. The form of magnesium called magnesium citrate is also a laxative and can help relieve some constipation that can arise during the luteal phase.

Vitamin E at 400 international units (IU) has some positive research for mood and breast pain symptoms, zinc at anywhere from 30 to 200 milligrams (which is quite a wide range!) for all PMS symptoms, and vitamin B6 at 40 milligrams for mood and physical symptoms of PMS. The quality of data is not robust, with variable doses and durations; however, similar to primary dysmenorrhea, there is a trend showing that lower vitamin, mineral, and antioxidant intake is associated with a higher incidence of PMS. This is another reason to work on diversifying nutrition and ensuring that your caloric needs are being met—if you're not eating enough, you're not getting enough nutrients either. A multivitamin may be something to discuss with your healthcare team, especially as you work on your nutrition.

In the botanical category, the champion herb is *Vitex agnus-castus*, commonly known as chaste tree. This is a big reproductive health herb traditionally and currently. It can help with many aspects of the luteal phase including premenstrual acne, breast tenderness, and spotting. The standard dosage is 40 milligrams, and the improvement in symptoms after using this herb for three to six months lasted for an additional three months once its use was discontinued. It's a fairly safe herb with the most common side-effect being a skin rash, which isn't all that common. It can be used in other forms (like tea and tincture) at different dosages. It has been used that way by herbalists historically, typically in combination with other herbs like chamomile.

Chamomile is a great option for those of you who might be experiencing anxiety and mood symptoms premenstrually (irritable, anyone?). As discussed in the dysmenorrhea chapter, chamomile contains compounds that exert anti-inflammatory, anti-spasmodic (i.e., anti-cramping), and anti-anxiety effects. As such, symptoms that benefit the most include cramping, anxiety, irritability, and breast pain. The dosage studied ranges from 100 to 400 milligrams in capsule form, twenty to thirty drops of liquid extract, or two cups of tea, and *all* of those were helpful to decrease PMS–even the tea–which is great! Teas are gentle ways of getting herbs in and a chamomile tea can offer complementary support for PMS (and a few moments of mindfulness). I'm a big fan of this for adjunctive support. For others who need higher dosing or dislike the taste of chamomile tea, capsules or extracts might be better options.

Other herbs supportive of stress and sleep can be used here as well, like lavender, passionflower, and lemon balm.

This is a good place to remind you that just because all of these things can help with PMS, it doesn't mean we employ *all* of these things for *your* PMS. You can use different modalities, but we need to take into account who you are and where you're at.

CHAPTER HIGHLIGHTS

- PMS occurs in the luteal phase and resolves once a period starts. It involves symptoms that are physical, somatic, and psychological, like bloating, acne, headaches, migraines, pain, breast tenderness, mood swings, nausea, fatigue, irritability, and insomnia.
- It is diagnosed when there are at least three days premenstrually where family life, social life, and the ability to work are impaired or family life and social life are impaired for five days over a minimum of two cycles. Your cycle tracking data is really helpful here.
- We don't fully understand PMS, but we do know that there is a hormonal change that impacts our body and mind. Early childhood and puberty-time trauma and stress are correlated with more PMS.
- There is no diagnostic blood test for PMS. However, relevant blood work for PMS can include CBC, ferritin, B12, vitamin D, thyroid panel, prolactin, and other parameters relevant to you. Vitamin D deficiency is related to a higher degree of PMS.
- Non-pharmacological interventions should be first-line treatment. These include counselling, nutrition, exercise, social support, adequate rest, heat applications, vitamins, supplements, acupuncture, and cognitive behavioural therapy.
- Talking about PMS with the people you live with, bonding with other menstruators over PMS, shifting your language to more neutral and positive phrases, working on self-care and stress management, and pacing yourself over your whole cycle, in both the follicular and luteal phases, can help ameliorate PMS (and the time available for self-care needed at that time).

- Under-eating, restricting carbs, not eating enough fruits and vegetables, consuming alcohol, too much coffee, and having high amounts of saturated fats and refined sugar can all worsen PMS. Focus on getting plentiful fibre through plant foods, enjoying antioxidant-rich fruits, eating regularly, incorporating carbs, having some omega-3-rich foods like chia seeds, and consuming soy to start shifting your PMS for the better.
- Calcium and vitamin E supplementation can help improve PMS, as can herbs like chaste tree and chamomile.

CHAPTER 17
PMDD

PMDD, premenstrual dysphoric disorder, is a diagnosis that was grouped under PMS for a long time. PMDD is not the same as PMS.

We're finally coming to understand it has a different mechanism even though it presents in the luteal phase as a result of hormone changes.

It is now included in the Diagnostic and Statistical Manual of Mental Disorders (DSM) as a mental health disorder. It impacts anywhere from 2 to 8 percent of menstruators globally; however, researchers surmise that this is an underestimation due to the lumping of PMDD with PMS, lack of awareness of the condition, and underdiagnosis of PMDD.

PMDD Assessment and Diagnosis

For a PMDD diagnosis,[8] five of the following eleven symptoms must be present, with at least one of the first four symptoms:

1. Markedly depressed mood, feelings of hopelessness, or self-deprecating thoughts
2. Marked anxiety, tension, feelings of being "keyed up" or "on edge"
3. Marked affective lability
4. Persistent and marked anger or irritability or increased interpersonal conflicts

---

[8] *Eisenlohr-Moul et al, The American Journal of Psychiatry, 2017.*

5. Decreased interest in usual activities (e.g., work, school, friends, and hobbies)
6. Subjective sense of difficulty in concentrating
7. Lethargy, easy fatigability, or marked lack of energy
8. Marked change in appetite, overeating, or specific food cravings
9. Hypersomnia or insomnia
10. A subjective sense of being overwhelmed or out of control
11. Other physical symptoms, such as breast tenderness or swelling, headaches, joint or muscle pain, a sensation of bloating, or weight gain

These symptoms are so severe that they interfere significantly with social, occupational, sexual, or scholastic functioning. They must occur in the second half of the menstrual cycle and resolve with the onset of a period, be present for at least two consecutive menstrual cycles, and not be a premenstrual exacerbation (PME) of an underlying medical condition or mental health disorder. This is all the more difficult because PMDD is associated with other mental health disorders such as bipolar disorder. Moreover, it is often misdiagnosed as another mental health disorder; once another diagnosis is in the mix, the likelihood of a PMDD diagnosis on top of that or in lieu of that is difficult.

BOX 10. Premenstrual Exacerbation (PME)

Premenstrual disorders encompass both, PMDD and PME. If you have an existing condition and experience a worsening of that this is called PME, or premenstrual exacerbation.

For example, if Zara has a diagnosis of major depressive disorder, and her depression gets worse after ovulation, then that's likely PME. This could look like her primary symptoms getting worse, symptoms associated with major depressive disorder that she usually doesn't have show up, or her usual management strategies not being (as) effective as they usually are.

Similar to PMS and PMDD, we need you to track your symptoms and listen to you as there is no diagnostic blood test for it.

Oftentimes, those with PMDD have a marked decrease in emotional regulation, cognitive function, and executive functioning in the luteal phase. This may manifest as extreme irritability, pervasive thoughts, volatile mood swings, lack of concentration, poor working memory, impulsive behaviours, and even self-harm and suicidal ideation.

People with PMDD often say things like:

- It feels like I have no control over my thoughts and emotions
- I feel like a completely different person
- I'm awful to my family and I feel like a shitty partner/parent/person
- I would use any measure necessary to not feel the way I do
- I honestly feel like I don't want to be around

PMDD awareness is still low. Family doctors sometimes do not recognize it or understand the extent of disruption and even suicidal attempts that can be part of the picture. Due to the volatility and distress involved in PMDD, it often needs more than one healthcare professional to support its different aspects.

I have learned the most about PMDD through patients sitting in front of me who struggle with all aspects of their lives in the luteal phase. It is a challenging condition.

We don't fully understand it yet although we're inching forward year by year.

In the past, PMDD had been related to a progesterone-deficient state; however, newer research shows a link between PMDD and a progesterone-derived neuroactive steroid called allopregnanolone and the GABA-A receptors, which are the same receptors that alcohol and benzodiazepine sleep medications affect to relax the nervous system; this GABA response seems to be diminished in those with PMDD.

Most hypotheses concerning the onset and experience of PMDD involve our stress response through the HPA (hypothalamic-pituitary-adrenal) axis and brain chemistry.

Other hypotheses to explain the mechanism behind PMDD include a different sensitivity to normal fluctuations in the HPA axis, dysregulation of the HPA axis, dysfunction of the serotonin system (serotonin is the neurotransmitter that is involved with feelings of well-being), genetics and epigenetics, alterations in calcium homeostasis, brain signalling modifications, grey matter changes in the brain, immune dysfunction, issues with histamine pathways, and/or shifts in the circadian rhythm.

It's interesting to note that in studies done in women with PMDD, early luteal phase estrogen has been lower and progesterone high in contrast–we're not sure what exactly to do with this information yet, but it does provide a clue that perhaps this relative hormonal ratio is playing a role.

There is an association between post-traumatic stress disorder (PTSD), childhood abuse, and sexual trauma and the incidence of PMDD.

Similar to PMS, PMDD risk factors include past traumatic events, being a competitive athlete in adolescence (which can influence stress and energy availability), loss of a close loved one at an early age, history of childhood neglect, and childhood physical or emotional abuse. This is in line with the HPA axis and brain chemistry theories on PMDD.

There is probably a role for intergenerational trauma here as well. We don't have much data on the incidence of PMDD after natural disasters or war, but it's not surprising when (racialized or marginalized) patients who have experienced significant distress or are descendants of those who have been displaced and traumatized come in with these conditions. (It's not surprising that there's a lack of this type of research in these populations either.)

There is some research on Japanese menstruators after the Great East Japan earthquake who had increased PMS and PMDD (as well as PTSD). We'll likely continue to see this play out in coming years with geopolitical conflicts globally.

Other factors associated with PMDD include cigarette smoking in adolescence and having a high BMI (which could be indicative of other medical, lifestyle, and socioeconomic variables and barriers to access and care).

Daily cycle tracking for two cycles is recommended to make a diagnosis of PMDD and to have a clear picture of what's happening. You can use a consistent scale of your own to monitor how you're feeling; however, there are tools designed to diagnose and monitor PMDD clinically. These include the Premenstrual Symptom Screening Tool (PSST), Calendar of Premenstrual Experiences (COPE), and the Dairy Record of Severity of Problems (DRSP), all of which allow for the rating of symptoms (mood, cognitive, and physical). A diagnosis can be given provisionally before employing those.

Similar to PMS, there are no diagnostic tests for PMDD, but there is a relationship with vitamin D deficiency. Testing and appropriate supplementation to reach adequacy is a supportive measure that should not be missed. Iron deficiency, vitamin B12 deficiency, and thyroid issues can independently contribute to feeling unwell, so testing those and ensuring adequacy is appropriate as well.

The International Association for Premenstrual Disorders (IAPMD) is a great resource for patients and practitioners alike; they even offer support groups and appointment prep sheets to facilitate discussions with your healthcare provider(s). There is a link to their website in Appendix B.

Treatment Options

The American College of Obstetricians and Gynecologists (ACOG) finally issued updated practice guidelines for PMDD in December 2023 that include not only medications but also exercise and lifestyle strategies.

In PMDD, there is a tremendous amount of difficulty with emotional regulation, impulsivity, emotional volatility, and non-fatal suicide behaviours that include suicide attempts. There is a need to assess suicide risk before any treatment is initiated, especially if treatment is being delayed for any reason. I hope you disclose this to your healthcare team and support system(s) if applicable to you. We want you to get the support you need.

I had an idea of what PMDD looked like theoretically through research studies. But honestly, only through the experiences of my patients and by going through PMDD forums online have I reached a better understanding of just how vast an impact this condition has on people. I've also come to learn how much listening, validation, and even mild improvements can have far-reaching effects on people feeling better overall. This is especially true if patients have experienced a lack of support and are feeling hopeless about their condition.

Discussing this diagnosis with people you live with or loved ones can be an important step. PMDD can considerably affect relationships.

For people with PMDD, there can be distress on top of an already distressed state because of how they're treating loved ones and how they don't have any semblance of control over how they are acting.

For loved ones, it can be difficult to bear witness to the volatile and extreme nature of their beloved. Mood can also be contagious, leading to an escalation of interpersonal conflict. Communication and understanding are parts of caring and loving; this might also allow both parties to figure out what helps and what doesn't.

Historically, PMS and PMDD were grouped together in research studies. The severity and conditions are different, but there is also a bit of overlap. As such, there is some overlap in lifestyle medicine supports here.

- Nutrition is an important factor, and we need to make sure that people with PMDD are eating enough for their body's needs. When we feed symptomatic people adequate food in the luteal phase, their low mood and energy levels improve. We also need to make sure that macro- and micronutrient needs are also being met and omega-3s are being consumed, so eating a diverse group of foods regularly is of benefit. If you don't eat enough, you're not going to get enough nutrients either. A bit of planning in the follicular phase might be helpful to have easy-access snacks and food available.
- Exercise has been shown to improve cognitive, mood, and physical symptoms in the luteal phase. Exercise is an underutilized therapy for most mental health disorders. Engage in whichever kind of exercise that you can get done.
- For smokers, working on quitting smoking is of benefit.

- Sleep disturbance is common in PMDD, and improving sleep is crucial for overall well-being. This can be done using mindfulness techniques, sedating herbs like chamomile and passionflower, supplements like melatonin and magnesium, other sleep hygiene practices (discussed in Chapter 30), and/or meds.
- Managing stress, using coping techniques, having school or workplace allowances in place, and a support system are all imperative as well.
- Community-based and online support groups can be beneficial.

Cognitive behavioural therapy (CBT) is a first-line treatment option for PMDD. Other forms of therapy and counselling, especially ones that help process trauma, can be helpful too; however, CBT has been studied the most.

In conventional medicine, the other first-line treatment is an SSRI (selective serotonin-reuptake inhibitors, which are typically prescribed for anxiety and depression), especially sertraline. Instead of taking an SSRI daily, SSRIs can actually be taken cyclically, i.e., in the luteal phase only. This reduces the distressing side-effects like fatigue, weight gain, and libido loss typically associated with SSRI use, and the benefits of the drug are still reaped in the second half of the cycle.

Over 30 percent of women with PMDD are resistant to (certain) SSRIs, so a bit of trial and error might be needed. It's important to remember that SSRIs (and other drugs) have interactions with medications, alcohol, herbs, and supplements. Please practice caution and discuss what you're taking with a knowledgeable healthcare provider to avoid harm and minimize potential interference. Other medications that are used to treat PMDD include the oral contraceptive pill (ideally a kind with low progestin levels), estrogen, or GnRH (gonadotropin-releasing hormone) analogs.

There are non-pharmaceutical options available. Chaste tree, or *Vitex agnus-castus*, can be considered in some cases of PMDD for mood-related and physical symptoms. Its safety profile makes it a good option, especially for those who cannot tolerate SSRIs. The dosage is usually forty milligrams for at least three months.

Saffron, an herb used in Middle Eastern and South Asian cooking, in supplement form at a dose of fifteen milligrams twice daily might be an option as well for its antidepressant effects. The Traditional Chinese Medicine (TCM) herbal formula called Xiao Yao San can also help with sleep, mood, and physical symptoms associated with PMDD. There are other herbs and compounds, like phosphatidylserine, that are used for mental health disorders showing promise, so we'll have to wait for more clarity.

Acupuncture helps with so many menstrual (and non-menstrual!) conditions, and PMDD is no different. It can be an amazing adjunctive therapy for anyone who has access to it. It helps with all aspects of PMDD using Traditional Chinese Medicine (TCM) as well as Korean acupuncture techniques. With acupuncture, ongoing treatment is most likely needed, just like any other practice or therapy. Some small studies have shown over 50 percent improvement in anxiety and depression scores with biweekly acupuncture over two months.

If you suspect that you or someone else has PMDD, please do not delay getting treatment. A multimodal approach with therapy, lifestyle modifications, herbal therapies, community support, medication and/or acupuncture can be life-changing, and even life-saving.

CHAPTER HIGHLIGHTS

- PMDD stands for premenstrual dysphoric disorder. It involves debilitating mental-emotional and behavioural symptoms in the luteal phase in combination with somatic and physical manifestations of hormonal change. These symptoms significantly interfere with social, occupational, sexual, and/or scholastic functioning. They resolve with the onset of a period.
- Cycle tracking for at least two cycles is required for a PMDD diagnosis to be made.
- We don't fully understand PMDD, but it has been linked to hormone change, stress and trauma, and brain chemistry changes. A vitamin D deficiency is associated with PMDD.

- Cognitive behavioural therapy (CBT) is a first-line treatment option for PMDD. Adjunctive therapies can include adequate nutrition, exercise, quitting smoking, sleep support, nervous system support, stress management strategies, chaste tree, and acupuncture.
- SSRIs, which are drugs used for anxiety and depression, can be used cyclically to provide support while minimizing the side effects of these meds.
- PMDD is a debilitating condition that can include suicidal ideation. If you or a loved one has or suspects PMDD, prompt treatment can be life-altering, and life-saving.

CHAPTER 18
Abnormal Uterine Bleeding

Abnormal uterine bleeding (AUB) refers to any out-of-normal-limits bleeding from the uterus. This encompasses terms like menorrhagia (heavy menstrual bleeding) and metrorrhagia (bleeding between periods).

Abnormal uterine bleeding can refer to the amount, duration, frequency, and/or regularity of blood loss. It can be acute or chronic. Basically, you have abnormal uterine bleeding if:

- Your menstrual blood loss is less than five millilitres or more than eighty millilitres per period
- You bleed for more than seven to nine days during your period
- You're bleeding less than twenty-four days or more than thirty-eight days apart, and/or
- You've reached menopause and experience bleeding of any amount

Context, obviously, matters.

Abnormal uterine bleeding is oftentimes not reported to doctors because it's normalized. Around 30 percent of menstruators will experience some form of abnormal uterine bleeding during their lives, and part of this can be due to very natural physiological changes like those seen at puberty and in the menopause transition. Sometimes, it's simply an adaptation response to something that has happened in your body and life. That being said, just because these changes are natural does not mean you need to go without support. It also does not rule out a concurrent pathology.

It's important to discuss out-of-range as well as out-of-your-norm changes in bleeding patterns with your healthcare team to understand if further investigation is necessary. AUB is related to a poorer quality of life and more anxiety. Trying to pinpoint patterns, getting appropriate care, purchasing and using numerous period products, having unexpected bleeds, losing days at work or school, and buying iron supplements are a few factors that impact people with AUB.

The acronym used by the International Federation of Obstetrics and Gynecology (FIGO) to classify the causes of abnormal uterine bleeding is PALM-COEIN, where PALM is related to structural issues and COEIN is related to non-structural issues. This is what the acronym stands for:

- Polyp
- Adenomyosis
- Leiomyoma
- Malignancy and hyperplasia
- Coagulopathy
- Ovulatory dysfunction
- Endometrial disorders
- Iatrogenic
- Not otherwise classified

Treatment of abnormal uterine bleeding will depend on a thorough history and the underlying cause of the bleeding. Bleeding of non-uterine origin should also be considered here. This includes things like vulvar or vaginal growths, sexually transmitted and urinary tract infections, inflammatory bowel disease, and pregnancy complications like pregnancy loss or ectopic pregnancy.

Now, let's break down this PALM-COIEN acronym further.

Polyp

A polyp is a growth of tissue that is typically benign. There is a very small chance that it's precancerous. Polyps can be found in different organs, like the gut.

In AUB cases, we're referring to uterine polyps (i.e., growths in the uterus). Sometimes, polyps resolve on their own, other times, they can be treated (with hormones or alternative therapies) if symptomatic, and sometimes they are removed and biopsied.

Adenomyosis

Adenomyosis is a condition where glands that are found in the endometrium enter the muscular layer of the uterus called the myometrium. Adenomyosis used to be considered a subtype of endometriosis–discussed in Chapter 20–especially since these two conditions are often found together, but not anymore.

Similar to endometriosis, adenomyosis is associated with a higher risk of fibroids. Some of the changes seen in adenomyosis include increased conversion of testosterone to estrogen, lower response to progesterone, increased blood vessel production, and repeated tissue injury and repair, which brings a lot of inflammatory compounds to the uterus.

Adenomyosis is generally diagnosed using transvaginal ultrasound and/or MRI. Research has found that MRI leads to a correct diagnosis more frequently than ultrasound. That said, most guidelines recommend transvaginal ultrasound as a first-line investigation tool. Histology (actually taking a look at the tissue) is another way to diagnose but requires a tissue sample, aka a biopsy. Researchers are currently looking at molecules that could serve as biomarkers for adenomyosis so we can eliminate delays in diagnosis to get people the care they need and deserve by looking at blood markers. Hopefully, this research will yield some useful results soon!

Adenomyosis can be asymptomatic in up to 30 percent of cases, but it is associated with secondary dysmenorrhea as well as heavy menstrual bleeding, painful penetrative intercourse, chronic pelvic pain, an enlarged uterus, mid-cycle bleeding, infertility, and adverse pregnancy outcomes.

The people at higher risk for adenomyosis are those who have had abortions, C-sections, higher parity (i.e., more births), and uterine curettage (removal of tissue from the uterus) since these factors carry with them the possibility of a change in the junction between the endometrial and myometrial layers of the uterus. Age and estrogen seem to play a role here, with higher age and more or longer estrogen exposure correlating positively to the development of adenomyosis.

We don't fully understand this condition (I know, it's an unfortunate theme of this part of the book), but we do have some treatment options that we borrow from endometriosis like painkillers, anti-inflammatories, IUDs, and drugs that target hormone conversion. Researchers are finding that those with adenomyosis also seem to be resistant to progesterone but progestins like dienogest (brand name: Visanne) which is also prescribed for endometriosis are viable options.

Surgery is another course of action but a bit risky for those who wish to conceive in the future, as postoperative pregnancies have a high miscarriage rate and chance of uterine rupture. This all depends on a variety of factors including what kind of procedure was done and the extent of the adenomyosis.

In the naturopathic and integrative world, we also borrow some of our therapies from dysmenorrhea (Chapter 15) and endometriosis (Chapter 20). This includes dietary changes to decrease red meat and increase plant foods, herbs to decrease heavy bleeding and pain like ginger, boswellia, shepherd's purse and yarrow, lifestyle changes like stress management and sleep support, and therapies like TENS machine and acupuncture.

Leiomyoma

This is a fibroid, which is an extremely common benign growth that can be found in the uterus on its own or in clusters. They can be fairly small or grow to be bigger than a grapefruit. Heavy menstrual bleeding, pelvic pain and pressure, frequent urination, constipation, and bleeding outside of a period are common symptoms related to fibroids. These symptoms can range from mild to debilitating, but you could also be asymptomatic.

It is estimated that fibroids occur in up to 70 percent of those assigned female at birth by the time they reach menopause, and there's a higher incidence of them as we age.

As with most conditions we've discussed here, we're not completely sure why and how fibroids form; however, we have theories around hormonal changes, structural issues in the muscle layer of the uterus, genetic factors, and even environmental exposures to endocrine disruptors. Age, hormone dysregulation, and a high degree of visceral fat (the kind around the abdominal organs) are associated with fibroids as well.

Fibroids tend to grow in the first trimester of pregnancy, with variable change in the third trimester–same, smaller, or larger. Similar to endometriosis, fibroids are responsive to hormones, especially estrogen; however, estrogen doesn't cause them. And, like the other concerns covered in this part of the book, vitamin D deficiency is a player; it's associated with the development and recurrence of fibroids.

Apart from chronic symptoms, the presence of fibroids can impact fertility and pregnancy outcomes depending on not only number, size, and location, but also the factors that contribute to fibroids in the first place, like age. They are linked to infertility, recurrent pregnancy loss, preterm labour, postpartum hemorrhage, placental issues, and a higher risk of C-sections.

Fibroids are typically diagnosed using ultrasound, more specifically through a transvaginal ultrasound, although sonohysterography (usually shortened to sono) is a useful tool to distinguish the type of fibroid present. MRI is the most effective diagnostic tool, but not used routinely because of the prevalence, ease, and lower cost of doing ultrasounds instead.

To be frank, fibroids can be challenging to treat. Shrinking them is possible, which can also lead to an improvement of symptoms, but it's difficult to completely resolve them without surgery depending on size, location, and number.

In the case of infertility, treatment options are a bit limited, especially if time is of the essence. When fertility is not on the table, treatment is typically determined by the desired outcome with respect to symptoms and quality of life: cessation of heavy bleeding, pain relief, etc.

Surgery is a treatment option: either removing the whole uterus, a hysterectomy, which is considered the definitive treatment, or removing the fibroid(s), myomectomy. A hysterectomy is reserved for severe cases and is not a particularly feasible option for many people, especially if you want to bear a child in the future.

Myomectomy tends to be a more viable option, especially if fibroids are interfering with fertility outcomes; studies are showing that myomectomies decrease the rates of primary infertility (i.e., there's an improved rate of pregnancy success in those who have not conceived before) and they lower pregnancy loss rates.

Depending on the symptoms, uterine artery embolization and focused ultrasound surgery that uses MRI guidance can be considered. However, all of these come with pros and cons. Ultimately, treatment choices should be thoroughly discussed with your healthcare team considering your long-term and short-term goals as well as your age and impact on quality of life.

Other treatment options include painkillers, hormonal contraceptives like the pill and IUD, ulipristal acetate (the same med used as emergency contraception), and leuprolide (aka Lupron) to decrease the stimulating effect of hormones on the fibroid(s). Lupron has been shown to decrease fibroid growth; however, it leads to bone loss and has been recommended to be used for a maximum of six months by the American College of Obstetricians and Gynecologists. It can be a viable choice if you're waiting for surgery or another clinical path.

Apart from correcting a vitamin D deficiency, there is some evidence for dietary changes similar to those seen in endometriosis like less red meat, more fruits and veggies, more fibre, more antioxidants, and less alcohol.

There may be a role for green tea extract standardized for its constituent called EGCG (epigallocatechin gallate) and enzyme therapy using serrapeptase here as well, but we'll gain some more clarity around that in the future.

There are herbs and other therapies that can assist in treating the symptoms associated with fibroids, slow down the rate of growth, and help them decrease in size. Acupuncture is one of these. Emerging studies are corroborating some long-standing and emerging herbal medicine combinations like ginger, turmeric, burr-reed, and a Traditional Chinese Medicine combo called tumour shrinking decoction composed of fifteen herbs that include turmeric, cinnamon, and poria. The key to formulating a treatment plan, as always, is to discuss goals and options and then go from there.

Malignancy and Hyperplasia

This refers to cancerous growth (malignancy) and/or unregulated growth of the uterine lining (hyperplasia).

For example, endometrial hyperplasia can be seen in people with anovulatory menstrual cycles like in PCOS or in perimenopause who have a build-up of their lining if their cycles are quite far apart.

Treatment depends on what the diagnosis is and will probably require a specialist such as a gynaecologist, endocrinologist, and/or oncologist, along with your family doctor and allied healthcare team.

Coagulopathy

Coagulopathies are conditions where there is an issue with blood clotting like von Willebrand disease.

There may be other signs of a clotting disorder, and a workup should include a thorough assessment to screen for easy bruising, nosebleeds, gum bleeding, family history, postpartum hemorrhage, and difficulty controlling blood loss from an injury, surgery, or dental work.

Treatment depends on the kind of condition it is, typically under the care of a haematologist (blood specialist).

Ovulatory dysfunction

Issues with ovulation can be seen in PCOS, hyperprolactinaemia, and thyroid disorders. This can also include bleeding patterns seen around puberty and perimenopause and can lead to changes in length and/or volume of flow.

Assessment and treatment can be a collaborative effort between your family doctor, endocrinologist, and naturopathic doctor based on the underlying issue.

Endometrial disorders

This is a diagnosis of exclusion when we can't find anything else wrong, so we assume that there's an issue with the endometrium. In this case, a gynaecologist takes point but other members of your healthcare team can be involved as a work-up is conducted and treatment strategy is laid out.

Iatrogenic

Iatrogenic means that something happened due to a medical treatment or procedure. In the case of abnormal uterine bleeding, the likely culprits are warfarin, aspirin, hormonal interventions like the IUD, and surgery. Treatment may include a cessation of medication and is based on what's going on.

Not Otherwise Classified

These are the other causes of AUB outside of the ones already discussed. This includes pelvic inflammatory disease, chronic liver disease, physical trauma, inflammation of the endometrium called endometritis, as well as malformations in the vasculature of the uterus. As you can imagine, the treatment will vary depending on the cause. It could include antibiotics, specific probiotic strains, physical and manual therapies, acupuncture, herbs, supplements, medications, and/or surgery.

BOX 11. What if my bleeding has changed but it's within normal limits?

Even though the parameters of this diagnosis are clear (as they should be), this doesn't catch people whose patterns have changed significantly from their normal, i.e., the so-called in-betweeners.

What happens if you typically lose twenty millilitres of blood with each period but now are losing seventy millilitres with each period (which is more than a three-fold increase)? Or what happens if you typically have a three-day period but now are having a seven-day period?

I'm not talking about the occasional time that you're seeing these differences, but a new, persistent shift.

Sometimes people with changes that are significant but technically within range slip through the cracks. It's not that these changes are always sinister (in fact, these are examples of what we might see in the menopause transition), but they do warrant a proper investigation.

CHAPTER HIGHLIGHTS

- Abnormal uterine bleeding refers to any out-of-normal-limits bleeding from the uterus like heavy menstrual bleeding, scant flow, bleeding after you've reached menopause, and prolonged menstrual bleeding. It can pertain to the amount, duration, frequency, and/or regularity of blood loss.
- PALM-COIEN is an acronym to classify the causes of AUB, where PALM refers to structural issues, and COIEN refers to non-structural issues. It stands for polyp, adenomyosis, leiomyoma (aka fibroids), malignancy and hyperplasia, ovulatory dysfunction, endometrial disorders, iatrogenic, and not otherwise classified.
- The treatment for AUB depends on what is causing and/or contributing to it. A referral to a gynaecologist is prudent in cases of AUB, especially if a cause cannot be found. Collaborative healthcare

that includes your medical doctors and allied healthcare team can be very helpful. Bleeding of non-uterine origin should also be considered.

CHAPTER 19
PCOS

Polycystic ovarian, or ovary, syndrome is a condition affecting a huge portion of the menstruating population, over 10 percent. I'm one of these people.

The term polycystic ovary syndrome is a bit of a misnomer because not everyone with PCOS has polycystic ovaries. There has been much debate and discussion in the last couple of years about renaming PCOS. Honestly, I'd rather those resources, time, and effort be used toward improving diagnosis and treatment for PCOS patients instead of resources going toward changing the name.

BOX 12. What are polycystic ovaries?

Having polycystic ovaries (PCO or PCOM for polycystic ovarian morphology) is different from having ovarian cysts, which tend to be fluid-filled sacs commonly seen over the menstrual cycle (more on those in Chapter 21).

Polycystic ovaries are when there are twenty or more follicles in at least one ovary upon imaging, there is an increased ovarian volume of ten millilitres or higher, and/or there are over ten follicles per section visualized in at least one ovary for older tech. Measuring and counting is part of a sonographer's job.

These follicles are in a state of arrested growth, i.e., they're not getting to the ovulation stage because of hormonal and metabolic dysfunction.

These are parameters meant for adults, not adolescents. Using ultrasound to diagnose PCOS is not recommended in adolescents as polycystic ovaries can be part and parcel of the first few years of the menstrual cycle being established.

PCOS is a lifelong multifaceted and multi-organ condition, not only a reproductive disorder.

Like most chronic and complex conditions, there is no singular cause. There are genomic factors, *in utero* exposures, environmental exposures, pre-pubertal nutrition and lifestyle contributors, and how all of these interact with one another involved. I'm going to dive deeper into this background (and everything else PCOS-related) in my next book, *The PCOS Literacy Handbook.*

Scientists have discovered numerous genes that play a role in the development and presentation of PCOS. These genomic effects range from insulin signaling and satiety to how hormones are produced and their action on cells. PCOS involves metabolic dysfunction, the hypothalamic-pituitary-ovarian (HPO) axis, inflammation, and the nervous system.

As such, it can present with many different symptoms to varying degrees and have many downstream effects on health as well. There are a few comorbidities such as thyroid disorders, vitamin D deficiency, infertility or subfertility, non-alcoholic fatty liver disease (NAFLD), irritable bowel syndrome (IBS), sleep apnoea, and even endometrial cancer. This last one is often due to abnormal and uncontrolled cell growth in the endometrium when periods are infrequent or absent; estrogen exposure without progesterone and without menstrual bleeding can lead to this kind of cell growth. This is one of the many reasons why it's important for you to share with your doctor if the interval between your bleeds is long.

PCOS Assessment and Diagnosis

Diagnostic criteria for PCOS have been a topic of contention for decades. The Rotterdam criteria for the diagnosis of PCOS are the most widely accepted ones, with an International Evidence-based Guideline update in 2023 fine-tuning the diagnostic criteria and standards of care. You have to meet two or all three out of three of the following criteria:

1. Oligo- or anovulation. This refers to infrequent ovulation or a lack of ovulation and often looks like irregular or absent menstrual cycles.
2. Clinical and/or biochemical signs of hyperandrogenism. This means that there are either elevated blood androgen levels (like DHEA, testosterone, androstenedione), and/or manifestations of high androgens in the form of hirsutism, pattern hair loss (around the hairline and crown of the head), and/or acne.
3. Polycystic ovarian morphology on ultrasound or elevated anti-mullerian hormone (AMH) in adults. AMH is produced by the follicles or so-called cysts seen in PCOS.

In addition to these criteria, we have to rule out other conditions that might be causing these signs and symptoms. These include congenital adrenal hyperplasia, which is a rare condition affecting the adrenal gland; hyperprolactinemia where the lactation hormone prolactin is elevated; eating disorders where we can see loss of menstruation and polycystic ovaries; and functional hypothalamic amenorrhea, which is discussed in Chapter 21.

PCOS is usually characterized by insulin resistance, high androgens, and inflammation. All of these feed into one another. Insulin resistance was covered in Chapter 4.

Signs and symptoms of PCOS can include: lack of a period for months on end, irregular cycles, hirsutism (excess dark and coarse terminal hair growth on the face, nipples, and pubic area), acne (especially cystic acne that might be refractory to treatment), polycystic and/or enlarged ovaries, high body fat percentage, weight gain or difficulty losing weight, lethargy, acanthosis nigricans (darkening of the skin around skin folds), digestive issues, pattern hair loss, blood sugar dysregulation, insulin resistance, subfertility and/or infertility, pregnancy loss and complications, disordered eating, mood disorders, and more.

It is a condition that is associated with vitamin D deficiency, subclinical hypothyroidism as well as Hashimoto's disease (where there are elevated antibodies against the thyroid), anxiety, depression, IBS, gestational diabetes, obstructive sleep apnoea, and higher risk factors for cardiovascular disease.

After menarche, menstrual cycles can be irregular and anovulatory for a while–this can look like PCOS, especially since polycystic ovaries, pubertal hair growth, and acne might also be present. A thorough investigation and appropriate follow-up care, as always, is important.

PCOS can impact fertility at different stages: conception, carrying to term, and pregnancy outcomes. I'll summarize some of these here: anovulation means that there's no egg, irregular makes it difficult to pinpoint the fertile window, ovulation predictor kits can give false positives as baseline LH levels can be elevated in PCOS, comorbidities like thyroid disease can exert their influence, elevated androgens and insulin resistance can contribute, endometrial dysfunction can affect implantation, progesterone production can be affected, and there are increased risks of gestational diabetes, preterm birth, intrauterine growth restriction, pregnancy loss, and preeclampsia.

This doesn't mean that all people with PCOS will experience infertility. It also does not mean that we can't have an impact on fertility; we can mitigate many of these risks and manage many aspects of PCOS! There are many strategies, conventional and naturopathic, available to improve fertility while reducing the risk of pregnancy loss and other complications. Collaborative care–starting *before* pregnancy–is super helpful here (more on that later)!

Regardless of whether fertility is on the horizon for you, or not, you *always* deserve respectful healthcare.

PCOS patients form the largest population in my private practice. It isn't uncommon for me to hear from patients who don't want to get pregnant (now or ever) that they were dismissed after being told: "Come back when you want to get pregnant". This always makes my blood boil.

PCOS, having a healthy menstrual cycle, and metabolic health are not only about fertility; it's about your whole health and life. Period.

A PCOS work-up should include testing for the diagnostic criteria, including a thorough history, exploration of signs and symptoms, ultrasound to check the ovaries, and blood work that includes androgen testing, typically testosterone and DHEA-S. It should also include other parameters to rule out other conditions, tell us about co-morbidities, and address any nutrient deficiencies like CBC, ferritin, B12, vitamin D, cholesterol panel, liver enzymes, FSH, LH, prolactin, estradiol, progesterone, AMH, insulin, glucose, HbA1c, oral glucose tolerance test, CRP, and a full thyroid panel including thyroid antibodies. And, it should include a conversation around all aspects of life and health like stress, sleep, mental health, eating patterns, and impact on quality of life.

Pharmaceutical and Conventional Treatment Options

There are a few pharmacological options that are employed in people with PCOS. If pregnancy is not on the horizon (now or ever), then the go-to has historically been the combination birth control pill. Irregular cycles can be extremely distressing to those with PCOS who cannot pinpoint when or if ovulation is happening. Taking multiple pregnancy tests can get stressful and expensive, so contraceptive options can be of great utility here.

Apart from offering contraception, the pill has a few benefits in PCOS: it prevents the overgrowth of the endometrial lining which reduces the risk of developing endometrial cancer, it helps reduce the anxiety associated with irregular cycles and the possibility of unwanted pregnancy, and it addresses acne and hirsutism by decreasing circulating free testosterone levels.

Although the pill reduces endometrial cancer risk, there is a slightly higher risk (compared to non-PCOS people) of developing blood clots. The other issue with using the pill for PCOS is that it typically only works on reducing symptoms while you're on it, which means your acne and irregular periods may return once the pill is discontinued. There are times when the effects last, but this is usually due to PCOS-supportive measures (intentional or inadvertent) put in place while on the pill, or if you've used it long enough to reach the age-related normalization of some PCOS symptoms.

Like the pill, the hormonal IUS will offer some of the same benefits like preventing endometrial hyperplasia and pregnancy, but not much support for the androgenic manifestations.

If you haven't had a period for three to six months, a hormonal withdrawal bleed is a prudent option to prevent the overgrowth of the endometrial lining called endometrial hyperplasia. In this case, you take oral progesterone for seven to ten days. Vaginal progesterone can be used as well but is typically reserved if oral progesterone is unavailable. Within a week or so of stopping the progesterone, a withdrawal bleed will occur to shed the lining that grew in the last few months (in a mechanism similar to that of a birth control pill withdrawal bleed).

Other drug options for PCOS may include metformin–the diabetes drug–which improves ovulation rates, cycle regularity, and pregnancy rate while reducing the risk of pregnancy-induced hypertension and late miscarriage risk via improved insulin sensitivity. Metformin does come with side effects, with the primary one being digestive disturbance. It is used in the fertility sphere quite a lot due to the reasons listed above; however, emerging evidence is pointing to an association between metformin exposure *in utero* and children having higher postnatal growth as well as increased central adiposity at five to ten years of age compared to children who were not exposed to metformin during pregnancy. If you're struggling with fertility issues as someone with PCOS, a timely referral to a fertility clinic can be helpful. Naturopathic and integrative strategies together can improve fertility while decreasing the risk of pregnancy complications.

There are also other medications that target PCOS symptoms. These include antiandrogens like spironolactone for hirsutism, acne, and hair loss, as well as retinoic acid (Accutane) for acne, neither of which can be used if you're currently trying to conceive or planning to soon. Acne, hirsutism, and hair loss are facets of PCOS that thousands and thousands of dollars are spent on, from laser treatments and chemical peels to skincare and hair care products.

There's also a procedure called laparoscopic ovarian drilling (LOC) which is a surgical procedure involving a laser or needle to restore ovulation in infertility cases refractory to ovulatory medications and carries risks of adhesions, ovarian insufficiency, and other complications.

Research has shown a positive impact of 5 to 10 percent weight loss on ovulation rates, pregnancy, menstrual cycle regularity, and metabolic parameters like blood pressure. As such, diabetes medications that target blood sugar and facilitate weight loss like semaglutide (aka Ozempic, a trending med these days) and liraglutide are sometimes used as short-term support. These medications come with side effects such as increased heart rate, nausea, vomiting, diarrhea, constipation, headache, fatigue, and more severe ones like thyroid, gallbladder, and kidney problems.

BOX 13. Eat Less, Move More

The most common "medical advice" PCOS patients are given, even by endocrinologists with a wait time of over ninety days, is to: eat less and move more.

Weight loss can be part of a PCOS strategy, but not for everyone. It should definitely not be the *only* medical recommendation that is offered, especially in this nebulous and oversimplified way. Excess fat, especially around the organs, contributes to insulin resistance, cortisol dysregulation, more inflammation, and hormonal miscommunication.

Fat is only one aspect of the body; there are other metrics of health that can be improved upon. Global data shows that less than 25 percent of patients (under the instruction of a doctor and with frequent accountability appointments) can maintain any significant weight loss in the long run. That percentage from other sources is less than 10 percent if we expand our inclusion criteria to have more of the general population in it. Success rates are typically dependent on having adequate resources, time, money, and other support like check-ins with healthcare and fitness professionals. Essentially, weight loss is quite unsuccessful for most people in the long run.

Treatment needs to be patient-centred, as always.

Many PCOS patients face a tremendous amount of sizeism, which is rampant in the medical system generally. The (sometimes) well-intentioned and often cavalier but simultaneously reductionist advice from doctors does harm. It perpetuates patient beliefs around food restriction, over-exercising, and self-blame.

In a population with a high prevalence of disordered eating patterns, body dysmorphia, anxiety, and depression, this feeds the sense of failure and lack of hope many patients feel. These patients end up thinking that there is something wrong with them and that it's their fault that they have PCOS.

If your doctor is hyper-focused on fat loss, try to find a different doctor or vocalize that you'd like to choose a different strategy. Unfortunately, even in the 2020s, we have data showing that family doctors and specialists alike in North America feel unequipped to support PCOS patients, oftentimes due to a lack of time and lack of simple educational resources to bring them up-to-date.

As a person with PCOS and as a practitioner who treats primarily PCOS in clinical practice, we need to stop telling PCOS patients to shrink their nutrition, their bodies, and their lives. We get this messaging about toxic beauty standards and body ideals from all kinds of places already. We need to move away from that and move more toward health and sustainable practices rooted in supporting long-term health, joy, connection, and flexibility. If your relationship with food needs help, then layers of support like individual or group therapy, books, podcasts, and check-ins with understanding healthcare providers might be of benefit.

Nutrition and Dietary Strategies

If you turn to the online space to find answers to your dietary and lifestyle PCOS questions, you'll be hard-pressed to find consistency in what's out there. There are all kinds of remedies and diets targeting those with PCOS. Please take what you read online with a grain of salt because PCOS has a strong genetic component, and it's not your fault that you have it.

From keto and paleo to intermittent fasting and severe caloric restriction (and everything in between), there is a lot of contradictory information on diet and PCOS.

If we look at the research, we actually need to shift our focus to a more holistic approach: one that includes eating regularly, incorporates all macronutrients (yup, even carbohydrates), and consists of lots of fibre and plant foods (think fruits, veggies, legumes, etc.). The addition of soy protein, flax seeds, antioxidant-rich foods like pomegranate and berries, and healthy fats like olive oil and omega-3 fatty acids are supportive. Decreasing refined sugars as well as saturated and trans fat while opting for more lean proteins is helpful.

This goes for all PCOS patients regardless of body size. Please do not cut out food groups. You do not need to cut out carbs and you do not need to eliminate fruits. You need adequate fuel to have a healthy menstrual cycle (and to have enough energy to take care of yourself!).

Another layer to a nutrition regimen can be to decrease the consumption of advanced glycation end-products (AGEs) found in red meat, charred animal products, and a variety of other processed foods that impact inflammation and cardiovascular disease risk.

These are the kinds of dietary patterns that are recommended for insulin resistance, fatty liver disease, and cardiovascular health in general. You'll notice similar themes across the dietary recommendations for menstrual and hormone-based conditions in this book.

Last, but certainly not least, we cannot forget about adding joy and flexibility in! We have to create space for the foods that bring pleasure in life. These can include nutrient-dense foods, but also other foods like popcorn and brownies–two of my favourites. There is no singular food that causes your PCOS, and there is no specific food that needs to be removed completely from your diet unless you're allergic or sensitive to it in some way. The one exception for some of you might be dairy, which may aggravate androgenic acne, and can impede the smooth flow of Qi from a TCM perspective.

Exercise and Physical Modalities

Exercise and movement are important for overall metabolic and hormone health, and this is true in PCOS as well. We know that exercise helps improve cardiovascular risk, mood, lean mass (like bone and muscle), insulin sensitivity, cortisol rhythm, fatty liver, cholesterol, cognition, and more. We don't have a ton of studies or guidelines on PCOS and exercise specifically at this point. However, any kind of movement and any kind of exercise is good for you.

Breaking up your sedentary time with movement is important, increasing your daily step count is important, adding in different kinds of activity and intensities is important, and joyful movement is important too. I like to follow the general Canadian exercise guidelines of a minimum of 150 minutes of moderate-high intensity activity, both resistance and cardiovascular training, plus lots of non-exercise activity.

Studies have shown that a short walk of ten minutes at a moderate pace after meals helps reduce insulin and glucose spikes after meals, so this is a place many of you can start. Remember, this movement is not punishment for eating; it is simply a strategy like a medication or supplement.

Data shows that high-intensity interval training (HIIT) is beneficial for the cardiometabolic risks associated with PCOS. For any of you worried about exercise and cortisol levels, check out Box 28 in Chapter 30.

Overexercising, under-fuelling or caloric restriction, irregular eating, yo-yo dieting, high unmanaged stress, and such are going to impact cortisol more than exercise alone. Exercise actually helps with long-term cortisol regulation, especially as we get better at exercise and engage in it consistently. It also helps with the different facets of PCOS.

Herbs and Supplements

Vitamin D supplementation is often a piece of the PCOS puzzle (no surprise there!). Vitamin D deficiency and PCOS are correlated, especially with respect to symptoms of mood dysregulation, insulin resistance, cycle regularity, and pregnancy rates. Even though the reference range for vitamin D in Ontario is 75 to 250 nanomoles per litre (nmol/L), it is better to be above the bottom of the reference range to see some of the aforementioned benefits. For PCOS, especially in fertility studies, we see more positive outcomes when the level is above 100 or 110 nmol/L.

One of the supplements we have the most research for is a compound called myo-inositol. Myo-inositol helps improve ovulation rates, cycle length, ovarian estrogen, egg quality, metabolic parameters like total cholesterol, and overall hormonal profile–it aids in lowering testosterone and DHEA-S. Myo-inositol has some important jobs in the body: it facilitates the actions of follicle-stimulating hormone (FSH), insulin, and thyroid-stimulating hormone (TSH). It is involved in the maturation of follicles under the influence of FSH, insulin signalling to control blood sugar and getting glucose into cells, and thyroid function under the influence of TSH; this last point might be part of the reason we see comorbid thyroid issues in PCOS, but we'll gain more clarity with time.

As with any therapy, myo-inositol isn't for everyone. Some people experience digestive upset with it, which can be mitigated with a divided dose like adding it to a water bottle, while others are not very good at absorbing it. We make inositol in the body and ingest dietary inositol through citrus fruits, beans, animal sources, and other plant foods as well. The dosage studied for PCOS is two to four grams per day. It does take some time (over three months) to see results with this, so consistency is key.

Another supplement that has a decent amount of research for PCOS is an antioxidant called N-acetyl cysteine (NAC). NAC reduces androgens, helps with egg quality, improves hirsutism scores mildly, and lessens circulating insulin levels. It also has a few fertility-related benefits that include improving ovulation and pregnancy rates with ovulatory stimulant medications, decreasing the chance of ovarian hyperstimulation with those medications, increasing the number of follicles, and reducing pregnancy loss rates. NAC is typically taken in capsule form, with the standard dose being 1.2 to 1.8 grams per day. It's generally tolerated well, but can cause digestive upset and doesn't smell particularly nice (it's a sulphur-based compound!).

Berberine, an alkaloid chemical found in plants like Oregon grape and goldenseal, helps with insulin sensitivity by increasing insulin receptor expression, and it decreases androgens. It has been used in those with diabetes and cholesterol issues and has some studies that show promise for PCOS. Even though the online world keeps likening it to Ozempic, it is not the same as the drug and works differently. Berberine is usually dosed at 500 milligrams per day and has been shown to improve cholesterol, blood sugar, and hormonal parameters in PCOS. Side effects with berberine tend to be gut-related.

In herbal medicine as well as Chinese medicine, there are herbs and combinations that are beneficial for PCOS. Some of these herbs, like licorice root, white peony, chaste tree, and spearmint, have all shown improvements in PCOS in studies as well. Spearmint, even in tea form, has shown to be slightly helpful for hirsutism at two cups per day, while the other herbs help with menstrual cyclicity, ovulation rates, modulating stress and inflammatory responses, and improving the hormonal profiles observed.

Then, there are other supplements and antioxidant compounds like quercetin, resveratrol, and probiotics that are being studied for different aspects of PCOS.

Many herbs, supplements, and topicals can be used for some of the symptoms associated with PCOS like acne and hair loss.

For acne, skin care products containing retinol or retinol alternatives, hydroxy acids, and niacinamide can be used along with light therapy and professional treatments. Avoiding abrasive exfoliants and ensuring consistent use of SPF are also important aspects of acne care.

For pattern hair loss, a mix of oral and topical agents can be used. This can include pumpkin seed oil, rosemary oil, phototherapy, microneedling, and topical formulas containing melatonin and EGCG from green tea.

Lifestyle and Stress Management

Acupuncture, therapy or counselling, seeking treatment from a dermatologist or medical aesthetician, connecting with others with PCOS, working with a nutritionist, and using a combination of conventional and naturopathic care can all be of benefit are all important aspects of care.

PCOS is associated with mental health disorders, especially anxiety, depression, and eating disorders. There is also a high prevalence of body dysmorphia and body image dissatisfaction issues. These are areas that most likely require long-term professional support and community support. (It also requires collective dismantling of the various systems of oppression that uphold toxic beauty standards that make us feel less-than!)

PCOS doesn't disappear. You'll need to look at your care from a long-term perspective!

CHAPTER HIGHLIGHTS

- PCOS is a common endocrine disorder involving metabolic health, sex hormones, and the nervous system. It is characterized by androgen excess, oligo- or anovulation, and/or polycystic ovarian morphology.
- Inflammation, insulin resistance, and high androgens are all interconnected in PCOS. It is multifactorial in origin with genetic, epigenetic, developmental, environmental, and lifestyle components.
- PCOS is a lifelong condition. PCOS care should be offered to anyone and everyone with PCOS, regardless of body size and fertility goals.
- Bloodwork, imaging, and your signs and symptoms should be considered for a PCOS diagnosis, to check for comorbidities like thyroid disorders and vitamin D deficiency, monitor progress over time, and inform your treatment strategy.

- There are many ways and layers of support that can be added in for PCOS to improve ovulation rates, regulate menstrual cycles, decrease androgens, improve insulin resistance, and support fertility. These include vitamin D repletion, nutrition rooted in balance and flexibility, acupuncture, therapy, exercise, body composition change, herbal medicines, inositol, NAC, and berberine.
- Conventional medical options include the pill, blood sugar-regulating medications, and anti-androgens.
- Eat less and move more is not sound medical treatment for PCOS.
- Strategies for symptoms like acne and hair loss can include a combination of topical and internal care.

CHAPTER 20

Endometriosis

Endometriosis is a condition in which tissue similar to your endometrial lining grows outside of the uterus. The tissue, referred to as endometriosis lesions, can be present in a variety of different areas: around the ovaries, in the digestive tract, around the bladder, in the rectum, and even as far as the lungs.

Endometriosis often presents with any, some, or all of the following:

- period pain that may not be responsive to usual dysmenorrhea strategies
- heavy menstrual bleeding, sometimes accompanied by clots and/or other symptoms
- presence of endometriomas, which are cysts
- pain with penetrative vaginal intercourse (dyspareunia)
- chronic bloating
- lower back pain
- bleeding between periods, like mid-cycle spotting or for many days before a period
- painful urination (dysuria)
- painful bowel movements (dyschezia)
- infertility, sometimes this could fall under "unexplained infertility"

- chronic or non-cyclic pelvic pain

It's the most common cause of secondary dysmenorrhea (period pain due to an underlying condition), but not everyone with endometriosis experiences painful periods. In fact, some people have endometriosis that is asymptomatic or presents with symptoms associated with other parts of the body. This can include recurrent urinary tract symptoms, an irritable bowel syndrome (IBS) diagnosis, recurrent ovarian cysts, migraines, generalized inflammation, and more.

Endometriosis tissue responds to hormonal fluctuations, similar to the endometrial lining.

In the online world, sometimes things get lost in translation, and it's made to seem like estrogen is the cause of endometriosis. This is not the case. Estrogen does not cause it; endometriosis *responds* to estrogen similar to how your uterine lining responds to estrogen. This also means that endometriosis lesions may respond to endocrine disruptors that are estrogen-like as well as other hormones in the body.

The underlying mechanism of endometriosis is not fully understood. We do know that there's a big inflammatory component with systemic immunological changes implicated.

Research has shown that there is higher oxidative stress in those with endometriosis–essentially, there's an imbalance between the body's capacity to deal with oxidants or inflammatory compounds and the speed at which those compounds are being generated or introduced in the body.

Our immune cells play a big role in keeping cell growth and pro-inflammatory processes in check while cleaning up after cellular debris. In endometriosis, there is a slight loss of these immunological functions at multiple levels. These changes are seen not only in endometriotic lesions but also in the endometrium itself. The immune compounds called prostaglandins, leukotrienes, and histamine that we discussed in Chapter 3 are part of all of this as well as the cells that produce them. There are some links to our gut and reproductive microbiome too.

Now, these immunological changes aren't the result of a singular reason. Genetics, biology, and the environment all contribute.

Layered atop all of that is the chronic pain component of endometriosis, which we're coming to understand more over time. In chronic pain conditions, there seems to be a higher sensitivity to pain. What may feel like a soft touch to you may feel like a harsh scratch or more to someone with a pain disorder.

Pain signalling in the body may change as a result of exposure to pain over time, and vice versa! Simply put, chronic pain can cause more sensitivity to pain, which equals more pain. This means that intervening and disrupting those pain-signalling pathways should be a priority.

It is estimated that around 10 percent of menstruators of reproductive age are affected by endometriosis and that it impairs fertility in around 50 percent of people with the diagnosis. This includes recurrent pregnancy loss.

Fertility issues are slightly complex in endometriosis–they have been linked to the presence of adhesions, endometriosis lesions, scar tissue, a change in the follicle's environment and development, poor maturation of eggs, altered response to progesterone, microbiome issues, altered development of the embryo, implantation issues, and a high degree of endometrial inflammation.

The extent of endometriosis lesions is not necessarily correlated with fertility outcomes. What this means is that you could have a few endometriosis lesions that can impact fertility, or you could have a lot of lesions and not have fertility issues (and everything in between). Obviously, the location of the lesions will matter.

A comprehensive discussion, fertility work-up, and an action plan are prudent sooner rather than later if you have endometriosis or suspect it.

Risk factors for endometriosis include a family history of endometriosis, earlier age at menarche, intrauterine and early life exposure to certain compounds like hormone disruptors, high or low birth weight, and environmental exposures, amongst others.

We don't yet fully understand the etiology of endometriosis, but we're understanding more and more that multiple systems are involved, especially immunological dysfunction and inflammation. There's lots of research being undertaken in this field and I do not doubt that we'll understand endometriosis a bit better in the next decade.

Endometriosis Assessment and Diagnosis

The average time to get a diagnosis of endometriosis in Canada is around seven years. The normalization of period pain by menstruators and practitioners, poor assessments, untimely referrals to gynaecologists, lack of practitioner knowledge around assessing endometriosis, use of hormonal contraception to manage symptoms, heterogeneity in presentation, and laparoscopy being the gold standard for diagnosis all are contributing factors.

Laparoscopy, which is a minimally invasive surgery, can diagnose endometriosis and be used to treat this condition. It has been considered the gold standard for diagnosing endometriosis for a long time. Laparoscopy techniques have come a long way in the last couple of decades; there is better technology and training to excise tissue and remove adhesions without damaging organs, especially for fertility preservation. As mentioned earlier, endometriosis is associated with infertility, and those classified as unexplained infertility have a high incidence of undiagnosed endometriosis.

MRI is an effective tool for diagnosis as well. It is especially sensitive to picking up deep infiltrating endometriosis and can map out lesions, especially before surgery. Getting an MRI, especially here in Ontario, can take a long time. Patients are ending up out of country due to delays at every step of diagnosis and care.

Up until recently, ultrasound wasn't a contender for diagnosing endometriosis, apart from very obvious lesions like an endometrioma, which is an endometriotic cyst. Advances in ultrasound tech, advocacy, and awareness, are allowing researchers to fine-tune the art of imaging and train sonographers to better use their tool to map out endometriosis lesions. Mapping techniques allow for a better understanding of how someone's endometriosis is presenting and affecting them. It also allows for a more comprehensive discussion about treatment options and how to move forward.

The more we can improve upon and utilize non-invasive techniques to diagnose, the faster we'll be able to diagnose and treat patients, especially if we can map out lesions accurately.

BOX 14. What about blood markers to understand endometriosis?

There aren't any diagnostic blood markers of endometriosis currently. There is research underway that may give us an answer in the future. Biomarkers are not widely available to run by all healthcare providers, but I think we will see an increase in availability as we learn when and where they have the most utility. Emerging endometriosis biomarkers are the same markers present in ovarian cancer.

There is a biomarker called cancer antigen 125 (CA-125), which is an ovarian cancer marker, that may be elevated in people with endometriosis.

CA-125, however, is not diagnostic of endometriosis. This means that if your CA-125 is within the normal range, we can't rule out endometriosis diagnosis—as in, you may still have endometriosis. On the flip side, if your CA-125 is elevated, we can rule it in, meaning that you may have it and we should follow this thread.

There is a positive correlation between the level of CA-125 and the severity of endometriosis, i.e., the higher your CA-125, the worse the endometriosis. If a CA-125 reading comes back as elevated, an ultrasound is the next step to determine whether there is a tumour involved or not as this marker is, indeed, elevated in ovarian cancer. Even though it's not a diagnostic marker, it can be a helpful tool to escalate your health issues if it's positive to get the appropriate referral and treatment. Other biomarkers include HE4, ASPN, and CA72-4.

Then, there are other tests to look into to understand your case and symptoms. This can include your CBC, where platelets might be elevated; general inflammatory markers like ESR and CRP due to the inflammatory nature of this condition; ferritin, thyroid, and clotting factors which can be involved in heavy bleeding; and vitamin D, as a deficiency has been linked to this condition.

If there are digestive issues, then Coeliac testing and an irritable bowel syndrome (IBS) work-up should also be included since these conditions are associated with endometriosis. There is an overlap between IBS and endometriosis symptoms, so sometimes endometriosis gets missed if a diagnosis of IBS has already been established. Menstruators with endometriosis are three times as likely to have IBS compared to those without endometriosis, so a thorough investigation is warranted.

If you're a practitioner reading this, I urge you to take your dysmenorrhea patients a bit more seriously (if you're not already doing so). Focus on developing a better line of questioning, especially around the symptoms commonly seen in endometriosis, so that we can support patients better. The Endometriosis Network Canada has great patient- and practitioner-facing resources. Their website is listed in Appendix B.

Pharmaceutical and Conventional Treatment Options

The quality of life of endometriosis patients can suffer greatly because of the severity, frequency, and duration of symptoms. There are many other considerations as well: healthcare, not being heard or understood, delay in diagnosis, management strategies, loss of work or school or time with loved ones, fertility and family planning, mental health, finances, and so much more. The film, *Below The Belt*, sheds some light on patient experiences with debilitating endometriosis symptoms.

Treatment for endometriosis is a bit complex and a referral to a gynaecologist is important to review the diagnosis, treatment options, your goals, impact on quality of life (including fertility if that topic pertains to you), and prognosis whichever route you decide to go. Gathering a team for various aspects of this condition is a good strategy.

Laparoscopic surgical removal of endometriosis tissue is an option typically performed simultaneously with diagnostic or exploratory laparoscopy. Like most procedures, laparoscopy comes with its challenges. The main trouble, at least in Canada, is the time lag in getting a referral to a specialist with this expertise, all of whom are difficult to get in with. Then, there can be fertility implications if not performed carefully due to the presence of scar tissue, removal of reproductive organs, and/or damage to the reproductive tract. When going for a laparoscopy, a robust discussion about the procedure is paramount. This procedure can be transformative for some people, excising lesions and removing adhesions that can form between organs, improving quality of life significantly. It's important to remember that lesions can return, surgery might not fully alleviate your symptoms, and there can be repercussions from the surgery itself.

Even though excision surgery is sought out often, it isn't the ultimate or appropriate treatment for everyone with endometriosis. We need to cater your plan to you.

Pharmacological treatment options for endometriosis can include oral contraceptives, which will shut off the hormonal system. It seems to be effective early on in the disease process, especially if contraception is desired as well. The use of progestins is the first-line pharmacological treatment with dienogest, or Visanne, being top of the line to decrease period cramps, pain with intercourse, and pelvic pain; it may even shrink endometriotic lesions. It doesn't lead to as much bone loss as some other progestins.

The hormonal IUS can be a good option for those who have heavy bleeds as well as pelvic pain. NSAID painkillers like ibuprofen, but more usually like naproxen and mefenamic acid, are often used in endometriosis, especially before a diagnosis is reached, or if hormonal options are contraindicated or tolerated poorly. NSAID use, as discussed earlier, can lead to digestive issues to the extent of stomach ulcers and colitis.

There are other medications on the scene, and these focus mainly on lowering estrogen to decrease the response in endometriosis lesions. These include drugs that fall in the group called GnRH analogues like Lupron and can lead to bone loss.

I think we're going to see more immune-based therapies for endometriosis moving forward as we understand the disease process a bit better.

Nutrition and Dietary Strategies

Regardless of what role conventional medicine treatments are playing in your endometriosis care–and I do highly encourage you to see a gynaecologist to explore your options–there is a role for support with natural and adjunctive therapies considering the immunological and inflammatory elements of this condition.

The dietary theme here is to reduce saturated fat while increasing dietary plants and omega-3s. Polyphenol antioxidant compounds from plants are being studied in isolation for endometriosis and may be one of the reasons why fruits and vegetables are so helpful, along with fibre and minerals.

Dietary patterns associated with the development of endometriosis include high red meat consumption, low omega-3 intake, low fruit and vegetable intake, and high trans-fat consumption.

On the flip side, we see decreased severity of pain when red meat consumption is limited or eliminated, and fruits and veggies are added in. These are strategies that are accessible to many of us and will not only improve endometriosis but cardiovascular and overall health as well.

Dairy does not need to be avoided unless you're sensitive to it or avoid it for other reasons, as those who consume dairy regularly have been found to have a lower risk of developing endometriosis. Caffeine and alcohol, similar to primary dysmenorrhea, may exacerbate endometriosis, so avoidance or cutting back may be part of your care plan.

Lifestyle, Stress Management, and Physical Modalities

Lifestyle-wise, sleep support and stress management are important for a multitude of reasons.

Oxidative stress (i.e., inflammatory burden) is elevated when we're not sleeping adequately and when we're physically and mentally stressed. Low pain tolerance and poor sleep tend to go hand in hand. Mental health can suffer with a new or existing diagnosis of endometriosis for a multitude of reasons including anticipatory pain, unknowns around family planning, health decision-making, and concurrent mood disorders. We want to, as much as possible, support you in your endometriosis journey, minimizing the negative influences and amplifying the positive impacts of the lifestyle factors we have some semblance of control over.

Lowering our exposure to environmental toxins and hormone disruptors is also prudent. Research indicates that we need to start paying attention to compounds such as PCBs (polychlorinated biphenyls) and dioxins, which can mimic estrogen and exacerbate symptoms. Studies have shown that this latter group–dioxins–may be linked to the development and propagation of endometriosis.

Dioxins bioaccumulate in animals people eat and are stored in fatty tissue, so avoiding saturated fat from animals is one of the ways you can decrease dioxin exposure. Since red meat and dairy are the primary ways people consume saturated fat, minimizing or eliminating red meat and perhaps opting for low-fat dairy could be practical ways of applying that information.

Physical and manual techniques can be of use in endometriosis, especially if you concomitantly have pelvic floor dysfunction, adhesions, a history of physical injuries, scarring, fibrosis, and/or abdominal surgery. This list includes the different therapies listed in the dysmenorrhea section, including pelvic physiotherapy, massage, chiropractic, and TENS. Acupuncture, especially with herbal moxibustion, can especially be helpful for overall quality of life improvement. All of these were discussed in Chapter 15.

Herbs and Supplements

Like all other aspects of endometriosis, we're still learning about these.

We can use our two previously discussed anti-inflammatory and pain-relieving herbs, ginger and boswellia., for pain and heavy bleeding. There aren't any clinical studies for endometriosis specifically (only animal ones so far), but we borrow from period pain studies (which we can assume had patients with–perhaps undiagnosed–endometriosis as part of the population included).

The herbs danshen (Chinese sage) and Scutellaria baicalensis (Chinese skullcap) also may have a supportive role here, according to animal studies and their traditional herbal use.

Vitamins C and E can help decrease pelvic pain, cramping, and pain with intercourse. These are fairly accessible, especially since the research done on those was at not super high doses–1000 milligrams and 800 to 1200 international units (IU), respectively. Biochemically, these make sense as well, when we review the roles of vitamins C and E in inflammation and immune function.

Melatonin is another supplement that is showing great promise in *in vitro* studies. In these studies, we see that melatonin can decrease the proliferative impacts of estrogen on endometriosis lesions, and it facilitates an improvement in egg quality. Hopefully, the next few years will shed more light on how useful it is for endometriosis.

Similar to primary dysmenorrhea, vitamin D deficiency has been linked with endometriosis. Hence, there is a role for testing and bringing low levels up to adequacy. Another compound is PEA (palmitoylethanolamide) which works by decreasing pain and neuroinflammation and has been showing promise for significant dysmenorrhea and chronic pelvic pain in micronized form.

Some other compounds and herbs are being studied for symptomatic relief and to help shrink lesions. The primary focus has been on plant antioxidant compounds called polyphenols, which can alter different aspects of the immunological processes involved like cell proliferation, cell adhesion, estrogen activity, and new blood vessel formation. Some of these are showing promise in *in vitro* and animal studies, like resveratrol, pycnogenol, curcumin, baicalein, and quercetin.

Therapies targeting the microbiome and gut-brain axis, including antibiotics, probiotics, and adjacent modalities are all being explored as well.

As always, please consult with a knowledgeable practitioner before you make changes to your care plan.

CHAPTER HIGHLIGHTS

- Endometriosis is a condition in which tissue similar to your endometrial lining grows outside of the endometrium. It's the most common cause of secondary dysmenorrhea (period pain due to an underlying condition).
- It can present with period pain, heavy menstrual bleeding, chronic bloating, lower back pain, mid-cycle bleeding, painful urination or bowel movements, infertility, and chronic pelvic pain. It is associated with IBS.
- The underlying mechanism of endometriosis is not fully understood. We do know that there are alterations in how the immune system is functioning and a large inflammatory component.
- Laparoscopy is the gold standard for diagnosing endometriosis and is often employed in a way that allows for simultaneous removal of the tissue. Ultrasound and MRI technologies are improving to allow for less invasive diagnosis.
- Collaborative care allows for a multimodal approach and informed decision-making. Hormonal options are available as well as painkillers. There is a role for diet and nutrition that is high in plant foods, low in trans and saturated fats, and high in omega-3s. Sleep, stress

management, acupuncture, vitamins C, D, and E, and anti-inflammatory compounds can be helpful.

CHAPTER 21
Other Menstrual & Hormonal Concerns

Functional Hypothalamic Amenorrhea

Amenorrhea is when periods disappear for three or more consecutive months.

Primary amenorrhea is when a person with a uterus and ovaries does not get their first period by the age of fifteen. This can be due to an anatomical issue, a genetic condition, or problems with the hypothalamus or pituitary glands.

Secondary amenorrhea is when a menstruator (who has been having periods) loses their period for three or more months; this can be due to thyroid issues, eating disorders, inadequate nutrition, rapid weight change, primary ovarian insufficiency (more on this later in the chapter), stress, physical trauma, or a pituitary gland issue.

For now, we're focusing on functional hypothalamic amenorrhea (FHA). The word "hypothalamic" here refers to a menstrual cycle block at the level of your hypothalamus in the brain, while the term "functional" denotes that the amenorrhea can be reversed once the physiological cause is addressed.

FHA is typically an adaptation response to some kind of stress–metabolic, physical, or psychological. Examples of such stress can include intentionally restricting food, not eating enough for your needs, overexercising (while not meeting energy requirements), chronic illness, emotional trauma, or physical injury like a concussion.

It is not associated with genetic conditions or another reproductive pathology. It may even be considered a natural variation in one's menstrual cycles to distribute resources effectively during times of stress.

FHA can happen at any age and is not contingent upon weight change. Even though it's commonly associated with a really low body fat percentage, it can happen regardless of body size (as can eating disorders). This is also part of a collection of symptoms called relative energy deficiency in sport (RED-S) in which athletes are not meeting their energy needs as a result of insufficient caloric intake and/or excessive energy expenditure.

The threshold at which a period will disappear is not uniform across all people– something that leads to FHA in me might not lead to it for you and vice versa. There is an association between FHA and prenatal stress exposure, early childhood experiences, stress around menarche, and epigenetics (i.e., how genes respond to your internal and external environment).

Essentially, what's happening in FHA is that the brain turns down reproductive activity as it detects a threat or significant lack of resources. As you can imagine, during times of high stress, survival, and not reproduction, is most important. Hence, the brain overrides those hormonal activities.

FHA involves our two main hypothalamic-pituitary axes: the HPO with the ovaries and the HPA with the adrenal glands.

If you think back to menstrual physiology covered earlier, gonadotropin-releasing hormone (GnRH) is the hormone the hypothalamus releases so that the pituitary gland can make follicle-stimulating hormone (FSH) and luteinizing hormone (LH). These two hormones then lead to the secretion of estrogen and progesterone.

In FHA, GnRH isn't released properly; its pulsatility is decreased, which in turn results in reduced LH, FSH, and estrogen. All of this results in a lack of ovulation and the subsequent disappearance of periods. There is other hormone involvement, including leptin, which is a hormone indicating adequate fuel intake, as well as kisspeptin–which is upstream of GnRH–and is being studied as a potential therapy for FHA.

Individuals with FHA, especially when longstanding and untreated, often have multiple areas of health that are affected such as low libido, poor vaginal lubrication, dysregulated stress responses, depression, and bone loss (aka osteoporosis). A bone mineral density scan and ensuring adequate nutrition, particularly calcium and vitamin D intake, are important considerations here if menstrual cycles have been missing for a long time. This becomes an especially high priority in teenage years as we put on most of our bone mass before the age of twenty years, with peak bone density at around age thirty.

An assessment for amenorrhea should be comprehensive. All the conditions that might be contributing to a lack of periods (for e.g., PCOS, pregnancy, pituitary gland tumour, primary ovarian insufficiency, structural blockage, etc.) need to be considered and ruled out.

Blood work and imaging should be requisitioned as part of the work-up as well as an assessment of mental health and a robust review of medical history, food intake, exercise practices, training volume, trauma, stress, sleep, drug use, lifestyle, and circumstances.

FHA presents with low levels of all menstrual cycle hormones because the nervous system has shut down the reproductive system to preserve other essential body functions. A progesterone challenge, where oral progesterone is given for seven to ten days, does not elicit a hormone withdrawal bleed because there's not much lining to shed in the first place. Once other causes have been ruled out, then the diagnosis of FHA can be given.

Broadly speaking, working toward regular cycles (and better overall health) can be a slow process when it comes to FHA–time, resources, and accountability are typical treatment requirements. There are usually layers of healing to unpack, and reactivation of the HPA and HPO axes can take months.

One's relationship with food, eating disorders, trauma, mental health, body dysmorphia, nutritional deficiencies, and overexercising usually require specific and professional support involving a team of healthcare providers. A multimodal approach works best and can include counselling, supplements, meal replacements, social support, and medications.

Ovarian Cysts

Ovarian cysts are fluid-filled sacs that can form on the ovaries. They are super common. Many of us have had them and may not even know that they were there since they typically resolve on their own.

Most ovarian cysts are benign and don't require any treatment or surgery; however, some can cause symptoms and even bleed or rupture, which does require specific care.

Certain factors increase the risk of developing ovarian cysts including medications like fertility drugs and tamoxifen, cigarette smoking, tubal ligation procedures, and hypothyroidism (underactive thyroid). Cysts can be sometimes found during pregnancy as well.

The most common cysts in our reproductive years are follicular and corpus luteal cysts. Follicular cysts are the result of a follicle that didn't rupture during ovulation either because there was too much stimulation of the follicle, or because there wasn't enough luteinizing hormone to release the egg. Corpus luteal cysts happen when the corpus luteum (which is the follicle shell leftover from ovulation) doesn't deteriorate like it normally does. These latter cysts are the norm during pregnancy because the corpus luteum sticks around to maintain a pregnancy.

Ovarian cysts can become haemorrhagic, i.e., bleed; however, typically they don't cause symptoms and resolve on their own. Even if they rupture, it doesn't always lead to pain; however, it can cause excruciating pain. I've had many patients head to the hospital for this reason.

Then, there are other cysts or cyst-like growths which can result from an overgrowth of cells like we see in cancer and endometriosis, or associated with another issue like an abscess or infection.

Most of the time, cysts are found incidentally on a routine or investigative ultrasound. Depending on the person, further testing might be indicated. Postmenopausal cysts are less common and should be appropriately evaluated for malignancy, typically in the form of a blood marker called CA-125.

One of the complications of having a cyst–and this will be accompanied by numerous symptoms like debilitating pain, nausea, and vomiting–is ovarian torsion. This is when the ovary twists. It can damage the ovary by cutting off its blood supply. This is a medical emergency and should be treated as such. The incidence of ovarian torsion is difficult to gauge, but it is more common in those with a cyst or mass in the ovary that is greater than five centimetres (around two inches).

As mentioned earlier, most functional cysts resolve on their own. Recurrent, persistent, large (greater than ten centimetres), or symptomatic cysts should be investigated as they may indicate an underlying issue that might be leading to their formation or malignancy. All of these factors along with menopausal status, age, and characteristics of the cyst will determine how to proceed. Too many times are patients brushed off when they have recurring or symptomatic cysts that are associated with another condition.

Precocious Puberty

Precocious puberty (PP) is essentially when puberty and secondary sexual characteristics (such as pubic hair, breasts, etc.) are present earlier than expected.

The diagnosis of precocious puberty is given when secondary characteristics develop before eight years of age in those assigned female at birth and nine years in those assigned male at birth.

Puberty is influenced by genetic, environmental, and lifestyle factors. Any and all of these factors can play roles in PP. There could even be a malignancy that leads to this.

Precocious puberty is further broken down into two categories:

- Central: This type is dependent on early activation of the HPO (hypothalamic-pituitary-ovarian) axis and involves GnRH (gonadotropin-releasing hormone).
- Peripheral: This type is less common and is independent of GnRH. It is associated with congenital adrenal hyperplasia (a genetic adrenal gland disorder), a tumour, or significant exposure to sex hormones.

An evaluation usually includes thorough history-taking, growth and development measurements, conducting a review of systems including neurological symptoms, discussing any recent trauma, infection, or potential hormone exposure, blood testing of hormones, bone age testing, and physical exam.

Management is typically undertaken by family doctors or pediatricians as well as pediatric endocrinologists. Treatment depends on what kind of precocious puberty it is, what some of the underlying factors are, how old the child is, and how the best short-term and long-term outcomes can be reached.

Primary Ovarian Insufficiency

Menopause is officially defined as the twelve months after the last ever menstrual period, with the average age somewhere around fifty-one years old. In primary ovarian insufficiency (POI), menopause is reached under the age of forty years old. It occurs in approximately 1 percent of menstruators.

POI used to be called premature ovarian failure or premature menopause. It is different from surgical menopause, which happens as a result of the removal of the ovaries.

POI is diagnosed when there are irregular or absent periods for at least four months, there are signs commonly seen in the menopause transition, and FSH (follicle-stimulating hormone) levels in the blood are elevated–over twenty-five international units per litre–on two occasions that are over four weeks apart. Research on biomarkers for detecting POI (earlier) is currently underway.

Other conditions should be ruled out. Typically, AMH (anti-mullerian hormone), which is produced by follicles and serves as a test for ovarian reserve, is low.

Consulting with a fertility specialist as soon as possible is recommended for anyone with fertility goals. There is a chance of natural conception depending on where along your journey you are; donor eggs can also be considered depending on the situation.

The theme of this part of the book is that we don't fully understand the causes of menstrual conditions, and POI is no different. There are certain genetic conditions linked to POI and other contributing factors can include environmental exposures, autoimmune conditions, family history, infections, certain medical treatments (in youth and adulthood), ionizing radiation exposure, metabolic issues, and more.

Similar to how the hormonal changes accompanying menopause naturally increase the risk of cardiovascular disease and bone loss, POI also comes with these risks. Since estrogen is being lost early, there needs to be a plan in place to minimize these risks.

The use of hormone replacement therapy (HRT) is strongly recommended, with the recommendation of having HRT in place until the approximate age of natural menopause in the population–fifty to fifty-two years of age. Hormone therapy may be continued beyond that depending on the case.

Counselling to support mental health can be extremely helpful as this is a significant life change and can come as a complete shock. Lifestyle modifications and healthy habits such as increasing weight-bearing exercise, eating lots of whole foods and fibre, quitting smoking, getting enough protein, stress management, getting restful sleep, and connecting with others with POI also have roles here.

Menstrual and hormonal concerns, as you might've noticed from these last seven chapters, are not well understood. They are multifactorial and complex. Treatment isn't so clear-cut either and should always involve a robust discussion, including when standards of care are being followed and when we're venturing a bit into theory.

As you move through the next part of the book, I hope you start to see that these conditions aren't only impacted by your genetics, nutrition, and lifestyle, but also where you live, how you were raised, and everything else. As the cliche goes: everything is connected.

CHAPTER HIGHLIGHTS

- Amenorrhea is when periods disappear for three or more months. Primary amenorrhea is when a person with a uterus and ovaries does not get their first period by the age of fifteen. Secondary amenorrhea is when a menstruator loses their period for three or more months.
- Hypothalamic amenorrhea is a shutdown of menses by the brain to conserve resources. There's a downregulation of reproductive function due to food restriction, over-exercising, trauma, and/or chronic illness. It can impact libido, vaginal lubrication, stress responses, mood, and bone health.
- Ovarian cysts are fluid-filled sacs that can form on the ovaries. They are extremely common, typically benign, usually go unnoticed, and tend to resolve on their own. Rarely, a cyst can cause ovarian torsion, which is a medical emergency. Recurrent and painful cysts might be associated with an underlying condition.
- Precocious puberty is when puberty and secondary sexual characteristics are reached earlier than expected–when secondary characteristics develop before eight years of age in those assigned female at birth and nine years in those assigned male at birth.
- In primary ovarian insufficiency (POI), menopause is reached before the age of forty years old. It occurs in around 1 percent of menstruators. It is diagnosed when there is oligo- or amenorrhea, FSH is elevated on two occasions, and low estrogen symptoms are present. When cycling hormones are lost earlier, the risk for bone loss and cardiovascular disease can go up, which is why hormone replacement

therapy and/or other supports should be considered as a diagnosis is made.

PART FIVE

PERIODS & THE WORLD AT LARGE

CHAPTER 22

Periods and Politics

Health is political; bodily autonomy is political; and medicine and medical care are definitely political. A book on periods that doesn't mention politics and systems of oppression, albeit briefly, isn't complete.

Factors such as historical bias in medical research, public health and policy, current affairs and geopolitics, the business of the healthcare and wellness industries, the erasure and appropriation of traditional forms of healing, wage and wealth gaps, and cis-heteropatriarchal capitalist systems focused on hyper-individualism have not only informed our health and healthcare but also how we relate to ourselves and the world at large.

These are topics we need to address to shed light on the inequitable access to care, our public health crisis (in Canada), the normalization of period problems, the stigma and lack of education on reproductive health, and how we go about creating change.

Sexism, racism, capitalism, classism, colonialism, ableism, ageism, sizeism, and other forms of oppression are woven into the fabric of our societies globally, keeping us busy, distracted, feeling inadequate, and overburdened. Oppression is about power and privilege. It is power and privilege that are at the root of our global crises like climate change, war, and genocide, our national crises like affordable housing, food security, and healthcare, and our local and personal ones like burnout, domestic violence, and trying to make ends meet.

Patriarchy and capitalism may look a bit different in different parts of the world, but the principles are the same–we have standards, ideals, and roles based on sex and gender rooted in scarcity and exploitation instead of abundance and nurturing. Androcentrism, which is having (cis-)males as the reference point, as the norm, and as the gold standard, is part of this as well.

These ubiquitous hierarchical constructs put limitations on how we think about life, what someone can or cannot achieve, and how we spend our time. They create divides between people, within ourselves as individuals, and with this planet we inhabit.

They impact standards of care in medicine, access to dignified healthcare, which policies are put forward including those on abortion and taxation, which diseases and populations we study and treat, and which lives matter and which do not. It is beyond time that we treat *everyone* with respect and dignity. We have made progress, but we still have a long way to go.

These factors influence you whether you're aware of it or not. They're part of the environment we've all been brought up in.

This is what these next few chapters are about. I'm not an expert in these areas; my understanding is limited and I'm only going to scratch the surface here. I'll highlight and address some history and research, discuss my observations and musings as food for thought, and (hopefully) plant seeds for us to move forward more holistically. I'll even share some of the paradoxes, cognitive dissonance, pessimism, optimism, rage, pride, and joy of being a healthcare provider that is part of both the medical and wellness industries, i.e., some of the contradictions and coexisting truths of being human!

The Social Determinants of Health

Your health started being shaped before you were even born–I don't only mean in the womb or with the egg and sperm you were made from, but also the generations that came before you.

It's also constantly being remodelled by your micro- and macro-environments, some of which have to do with the luck of the draw, like nationality, birthplace, and family of origin.

This is where the social determinants of health come into play.

The World Health Organization (WHO) defines the social determinants of health as the non-medical factors that influence the health and health outcomes of individuals and populations. These include education, food insecurity, income and social protection, working life conditions, unemployment and job insecurity, access to affordable and decent health services, housing, basic amenities and the environment, early childhood development, social inclusion and non-discrimination, structural conflict, and bias and discrimination based on race, sex, gender, resource availability, or other factors.[9]

We've addressed a few of these already throughout the book. They are part of the reason why health, health outcomes, and the management of health conditions can look massively different depending on who you are. Research shows that 30 to 55 percent of health outcomes are dependent on these social and *structural* elements.

They also highlight the inequities we see in health statuses within our nations as well as in relation to others.

There is a social gradient here: the lower the socioeconomic position, the worse the health (and privilege, opportunity, and resources). We can frame this in a slightly different way: the more exploited the nation, the worse the health (and privilege, opportunity, and resources) of the people who live there.

The people who tend to suffer the most are lower-income, racialized families, and other marginalized groups who get caught in a loop of low opportunities and do whatever they need to do to make ends meet.

These are the same people with histories of generational trauma, displacement, food insecurity or starvation, elevated exposures to environmental pollutants, a high degree of stress, and increased rates of mental health diagnoses and

---

[9] *World Health Organization, Social Determinants of Health, Updated 2023*

incarceration. These are also the same people who are profiled, stereotyped, and sometimes resort to desperate measures to make ends meet. This is *not* a failing on their part, it is a failing of the governments and policies that think of them as less than, if it all.

How can you receive comprehensive medical care when you live in an area without a doctor or emergency services? How do you eat nutrient-dense foods when you live in a food desert? How do you not miss school when you cannot afford appropriate period products? These are only a fraction of the things that many of us are lucky enough to not have to think about. These are also very real experiences of many marginalized folks.

Our governments and public policies perpetuate these inequities, disallowing huge pockets of our population from accessing basic securities that people with power and money have the most access to. Many of our politicians are corrupt and further corporate and selfish interests at a very real cost to the people they are supposed to serve.

Where I live, the provincial government had billions of dollars from pandemic relief efforts to spend on public healthcare that they *chose* not to spend to improve hospital wait times, living wages for nurses and medical personnel, and to make much-needed improvements to our public healthcare system. They continue to underfund healthcare.

What happens when you push something to its breaking point by depriving it of sustenance? Something has to give—patients slip through the cracks, nurses leave the profession, medical doctors leave for the private sector, and people suffer. You can also put forth so-called solutions that are self-serving.

Enter: a two-tier system; a system that widens the gap between the 1 percent and the rest of us. Canadian and foreign corporations have been lobbying for healthcare privatization for a long time. It is underway in Ontario. We have millions of people in the province without a family doctor. This puts a higher burden on every level of the healthcare system.

Universal healthcare is supposed to help *everyone*. Privatization does not. It puts the people who need care the most at the bottom of the list and in the hands of underfunded institutions. It also creates more burden, burnout, and life dissatisfaction in those working in the public system due to a lack of resources.

There are layers and layers to peel back here; however, we're not going to find our answers in corruption, profiteering, and short-sightedness.

I'm also not saying that there's an easy solution here. What we do need is more accountability, a focus on our pressing issues like climate change, the exploitation of our planet and its people, and better infrastructure that doesn't only benefit the few.

Capitalism and Collectivism

Capitalism puts profit ahead of people and our planet. Furthermore, there are no real repercussions for those who exploit and destroy the most; it's the rest of us who bear and will continue to bear the burden of reckless power plays, especially those who are marginalized the most and contribute to these issues the least.

Due to globalization, political and economic unrest in any part of the world will have a domino effect that trickles into our day-to-day lives. Between the COVID-19 pandemic and geopolitical conflicts, the early 2020s are an unfortunate but perfect example of this. Emotional turmoil, hate crimes, and the rising cost of living are only fractions of how all of this plays out at the individual or community level.

Why am I bringing all of this up?

Because everything is connected. Because we direct our solutions toward the individual when we need more accountability at the corporate and institutional level. Because these are the current frameworks that are in place. Because I want you to remember that your health and your life do not exist in a bubble and are heavily influenced by all of this. Because I want you to have some compassion for yourself and where you are in your health and life journey. And, because we need everyone to be part of the solution.

I have friends and colleagues scattered across different healthcare fields who say that they don't want to get political. I'll repeat what I wrote at the beginning of this chapter: health is political; bodily autonomy is political; and medicine and medical care are definitely political.

Discourse and training are scarce in topics related to these very real issues that the average person is dealing with. We need to include more discussion concerning the structural determinants of health, systems of oppression, and cultural sensitivity in institutions and medical systems, including in my own profession—a profession that is supposed to treat the whole person while identifying and treating the causes.

How our patients feel–and how we feel as doctors and humans–is not detached from what happens in our world.

Injustices around us–yes, the ones that impact us directly, but also those that do not–vastly influence our health and well-being. They affect our connection to ourselves and our purpose, our mental health and capacity to perform, and how we relate to the world and others. They help us identify the limits of what we can and cannot accomplish under the circumstances, understand that rage and heartache are normal reactions, let each other know that we're not alone, bolster one another, and take individual and collective action.

We have to take a critical look at our environments, our education systems, our media channels, and our policymakers. We have to stop morphing and shrinking ourselves to fit into the systems and, instead, disrupt the systems so that they're not only benefitting the few.

We need strategies that will decrease the wealth gap, stop the exploitation of our planet and its inhabitants, and redistribute wealth and resources to sectors and places that benefit more people, especially those who have typically been left out.

We need to work on livable wages, better education, improved access to nutritious food, affordable housing, comprehensive transportation infrastructure, more parks and green spaces, affordable childcare, more debt forgiveness and less predatory lending, improved healthcare access, acknowledging past harm, no- and low-cost community programming, patient-centred care, political transparency, and all the other supports that will *actually* help our communities.

We have to come together to innovate solutions and strategies while holding big corporations and responsible parties accountable. We don't have to reinvent the wheel: there are nations and groups we can learn and adapt from. These are the approaches that'll help more people, fostering joy, peace, community, and collaboration–the measures that will lead to lesser stress and better health.

What this also means is that we must learn to pause, rest, and take care of ourselves so that we have energy and space for creativity, innovation, imagination, and connection–that is how we reclaim our time and health. (And, that's the way to more peaceful periods as well, by the way!). As such, we have to start paying attention to our bodies, our thoughts, our intuition, and our traditions.

Working toward health equity means taking our whole population seriously. Humanity, health, well-being, and abundance are neither built upon the tearing down of other fellow beings nor the exploitation of this planet we call home–they are built by lifting each other and taking care of our home.

It can be very easy to lose hope and feel powerless. We've come a long way *and* we have a long way to go.

There are many people, communities, and organizations working hard to make change happen. I believe that we can all do something regardless of the perceived size of the action. You can contribute at the level of yourself, your family and friends, your local community, your national community, the government, political agencies, nonprofit organizations, and schools.

Just like the rainforest is reliant upon all parts of its whole, we need people in all kinds of places and all kinds of roles. We've been seeing the literal and metaphorical deforestation of our world, and the balance is askew.

If you have money, time, or other resources to offer, support people and organizations who are aligned with doing the work of creating a more equitable world (Appendix B has a few organizations). Engage in learning, discussion, art, and debate. Hold on to hope. We can all do something.

CHAPTER HIGHLIGHTS

- Research shows that 30 to 55 percent of health outcomes are dependent on the social determinants of health, which are non-medical factors that influence health and health outcomes of individuals and populations.
- The systems in place need to be disrupted as they are not designed to support the average person–they put profits, power, and privilege ahead of people and our planet.
- Our micro- and macro-environments affect our menstrual cycles and health. Health doesn't exist in a bubble. We need to continue to work on individual health and public health measures simultaneously. The way we help ourselves and others is the same: equitable measures and collective upliftment.

CHAPTER 23

Menstruation and the Medical Industry

Great strides have been made in medicine and healthcare. Technology and scientific research have allowed us to understand, diagnose, and treat a greater variety of health conditions than ever before.

As we rejoice in furthering our understanding, we must acknowledge the dark history of gynaecological research upon which our understanding rests, specifically the torture and forced experimentation and sterilization of black, Indigenous, and racialized folk that you can read about in books like *Medical Apartheid* by Harriet A. Washington. You can add to that the ongoing appropriation and erasure of traditional and cultural medicines amongst other atrocities in the name of science.

Scientific studies on menstruation and menstruators were far and few before the beginning of the twenty-first century. One significant contributing factor to this was the US Food & Drug Administration's exclusion of people of childbearing age in drug trials–for a reason. It was a response to the thalidomide debacle in which childbearing patients across Europe and Canada taking the drug resulted in offspring with severe deformities.

Menstruators have been historically excluded from research for other reasons as well. Menstrual research typically requires a longer time frame (and more resources) to account for the variations over the phases of the cycle and to conduct studies beyond a single menstrual cycle. Pressure from publications and institutions for quick turnarounds is common. Who conducts the study, the kind of research done, the procurement of funding including the amount of money acquired, how the results are utilized, and other facets of scientific research have been at the mercy of the people running the show.

There has been a general lack of care and regard for women, who continue to fight for equity and agency over their bodies to this day. There are additional layers here based on ethnicity, gender identity, sexual preference, body size, disability, social status, neurodivergence, and more.

Marginalized folk tend to have a deeper mistrust of the medical system, research, and government–for good reason. The prime era of eugenics wasn't all that long ago, and both xenophobia and white supremacy are well and alive today.

These are the same populations that also have the poorest access to dignified care and the highest amount of negative healthcare experiences. As such, these populations have historically been subject to the least amount of attention.

Research has never been as diverse as it is right now and that body of research continues to grow, which is great! We've extrapolated and applied research from (cis-white) males to other populations for a long time. That data has been and is valuable! It also comes with limitations.

Certain conditions are more prevalent in women, like autoimmunity. Hormone changes over the lifespan have a major impact on how our health journeys play out. We didn't assess women for cardiovascular disease as robustly in the past. The more research we do, the broader and more nuanced our understanding becomes.

Data on transgender and gender-diverse folk is sparse. We know that these populations tend to have a high prevalence of avoiding medical care for myriad reasons including feeling unvalidated and unsupported in their healthcare. Many of these folk experience dysphoria associated with puberty, are at high risk of social isolation and discrimination, and could use better care in general.

We also need more research on menstrual suppression at puberty and how we can best support patients with their goals and health.

Even where there is research, we are often limited by the model of car we deliver and the system this model is predicated upon.

The fact is: that our modern medical model and health industry is built upon and perpetuates systems of oppression like racism, sexism, ableism, sizeism, etc.

Amongst other harms like disrespect, medical trauma and the dismissal of pain and period problems, this has led to many standards of care and diagnostic tools that aren't inclusive. The under- and misdiagnosis of skin conditions in melanated people is a prime example of this. Medical education, including naturopathic education, primarily uses white skin to teach pattern recognition in dermatological disorders, which can present differently on darker skin. These are the kinds of teachings internalized by practitioners. We don't even think to look or go beyond what we've been taught, especially when students and practitioners alike are part of an education model that is draining physically, mentally, emotionally, spiritually, and financially. I can easily relate to that myself; I didn't think much about how guidelines are developed, what and who is missed, and how this impacts the health and life of real people. It's a sad–but not unsurprising or unexpected–reality.

We see, time and time again, androcentrism as the gold standard as well as capitalism in the business of healthcare. Even in public systems, like where I live, profiteering triumphs over public service. In fact, private interests lobby the government to underfund social support causing public dollars to flow back into these private companies which feeds back into more of the same.

As mentioned in Chapter 22, the biggest barrier we face is systemic–these capitalist models that don't care for the average human being. We need governments that care about people, the planet, and the future–not simply look at humans and Earth as things to exploit. We need better infrastructure to help a larger number of people. There are ways to do this, but it requires petty and power politics to be set aside.

The wellness industry isn't all that different, especially in its latest iterations. From cultural appropriation and over-pathologizing regular human variations to predatory marketing and disregard for public health measures, the Big Wellness is predicated on the same principles as described above.

Celebrities and personal brands are being used to capitalize on foods, anti-aging formulas, hacks, courses, supplements, retreats, spas, lifestyles, and gadgets. We've commodified wellness and we've done it spectacularly. The wellness economy has grown to trillions of dollars.

It has us chasing body and life ideals that seem to be ever-evolving and ever-elusive instead of focusing on what really matters: being here, loving hard, taking care of ourselves and loved ones, and finding joy in our day-to-day life. Yes, some of it is rooted in good intention, and some of it is rooted in truth, tradition, and science; however, mis- and disinformation are also rife, as is the wanting of "more" and "better". From twenty-step skincare routines to thirty supplements per day and everything in between, we've become self-improvement projects like never before.

I'm not trying to downplay the very real successes and advances we've had in health and life thanks to the medical and wellness industries–I love that we have a better understanding of and a larger breadth of tools to facilitate health and well-being than ever before.

I'm also not trying to discount the very real benefits of preventive and integrative medicine, implementing healthy habits, taking care of your mind, body, and soul, or feeling well in your body and life. I want people to feel well and live the best life possible! Heck, I am a regulated health professional with my feet in both, the medical *and* wellness industries. I even wrote a whole section at the end of this book on hormonal health promotion and menstrual wellness.

I work within the confines of all of the parameters I've laid out while having a love-hate (and all the other feelings!) kind of relationship with them. I know I'm not alone in that. All of these truths can and do coexist.

I simply want you (and myself) to focus on the foundations first, access medical care, understand that collective health and individual health are intertwined, and use discernment as you do all of that. I want you to know that there's a difference between medical management of a chronic condition and implementing habits for prevention or general wellness.

I want us all to ask ourselves why we're chasing a particular thing, what we want to feel and do in life, who we're looking up to, where we're getting our information, and how we spend our time and resources.

I want us to be well in more holistic ways, in our own personal ways, and in real ways that foster and build a robust foundation for health, self-love, friendship, community, and collective upliftment. Many of us have the privilege to be able to choose what health and well-being are to us. Don't go chasing someone else's definition of it. You do you.

BOX 15. Health versus Wellness

The terms health and wellness are often used interchangeably. As you reflect on those terms and think about what they mean to you, I'll share some existing definitions to help you along the way.

The World Health Organization (WHO) defines health as: "a state of complete physical, mental and social well-being and not merely the absence of disease or infirmity." [10]

It's not too dissimilar to the National Wellness Institute's definition of wellness, which is:

"Wellness is functioning optimally within your current environment. Wellness is a conscious, self-directed, and evolving process of achieving one's full potential. Wellness encompasses lifestyle, mental and spiritual well-being, and the environment. Wellness is positive, affirming, and

[10] *World Health Organization, Constitution, Updated 2023*

contributes to living a long and healthy life. Wellness is multicultural and holistic, involving multiple dimensions."[11]

I'm going to leave that there for you to chew on. I do think having some clarity in what health and well-being mean to you as an individual helps you be true to yourself and how you want to live.

By all means, take inspiration from others; however, don't get caught up in someone else's ideas of wellness. I can confidently share that my definition and relationship with the concepts of health and well-being are constantly evolving; over time, there's more nuance, clarity, and simplicity.

Returning to the topic of medical research and healthcare, we are seeing tremendous change. Thanks to the advocacy and work of many individuals and groups across the globe, the research landscape, health conversations, and medical practice have been changing (hurray!).

There's more (open) discourse on historical oversights and exclusions, equity and inclusivity, social justice, reconciliation, the importance of individual *and* collective care, honouring traditional and cultural practices, decolonizing medicine and therapy, and planetary health.

It can be difficult to capture the nuanced complexities of health, especially those about the interconnected and multilayered roles of the social determinants of health. Even when research exists, interpreting and synthesizing all of this information for practical use can be challenging for a variety of dynamic reasons. That being said, we're moving in the right direction.

We're seeing an increase in menstrual research being available to us in diverse populations in the fields of physiology, diagnosis, biochemistry, pharmacology, nutraceuticals, nutrition, epidemiology, policy, product development, education, innovation, and integrative therapies.

All of this will hopefully help us move forward in providing more wholesome care to those who need it, but also in affecting policy and institutional reform

---

[11] *National Wellness Institute, NWI's Six Dimensions of Wellness, 2023.*

that is necessary to support the average person who is struggling to meet the basic needs of their family unit.

The trickling down of change into medical practice is often slow, though. Many barriers exist to the accessibility of patient-centred, safe, respectful, collaborative, and evidence-based care when it comes to menstrual and hormonal health.

There is a lack of training and discourse in medical education on compassionate care, cultural sensitivity, the history of medicine, the role of doctors as teachers, and the intersectionality of health. Additionally, there is a lack of focus on important concepts such as informed consent and shared decision-making. Discrimination such as sizeism, racism, ageism, ableism, and gender bias are still rife. Then, there's also implicit bias, personal limitations, and ego that can get in the way. If I look at my naturopathic medical education, which is based on treating the whole person, we had very little training and discourse exploring these areas.

To top it all off, practitioner and health personnel burnout is extremely common as there is an enormous burden of responsibility that is inherent to the job as well as training for the job.

Medical practitioners are humans, just like everyone else, at the end of the day. Yes, individual work matters, and I encourage any healthcare practitioners reading this to critically reflect on your personal and professional lives and take action to be better. At the same time, we need change at all levels, from schooling to improving support for practitioners through easy resources and better healthcare funding.

We need strategies to reach more people and to make it easier for medical practice to evolve. Simple educational tools such as video materials and small group discussions help healthcare providers check some of their biases and shift their behaviours. These are low-cost and easily implementable.

There are many other ways and much room for growth. Policies and practices that involve a collaboration of people like scientists and researchers, healthcare providers, patients, educators and institutions, policymakers, community representatives, and other organizations offering input and actionable items will hopefully affect change that is meaningful and impactful by addressing the multitude of factors that make up our health.

Your Menstrual Health and Medical Care

In Canada, the family doctor is the gatekeeper to all specialists. Depending on your doctor's knowledge, experience, diagnostic skills, understanding of their limitations, and consideration for your experience and story, you may or may not (a) receive appropriate care, and (b) be referred to a specialist.

The thing about many menstrual and hormonal health concerns is that there are crossover elements. PMDD and perimenopausal changes are good examples to highlight this: there are often hormone-based changes with psychological and physical manifestations that span a few physiological systems. This poses an interesting problem: whose job is it to support patients anyway?

Another obstacle we have faced for a long time–and continue to face–is the normalization of menstrual health issues, especially the most prevalent ones of PMS and period pain. Unfortunately, this is by menstruators and practitioners alike. The education system and our medical practitioners don't really teach us much about menstrual cycles or when to ask for help, which gets in the way of patients being able to access the care they need and deserve.

Moreover, in conditions like PMS and pain, where often there is no visible pathology or diagnostic blood test, the validity of health concerns can be questioned. Practitioners need to do a better job at listening to patients and using the diagnostic criteria and tools we have at our disposal. My colleague, Dr. Kara Dionisio, ND, said it beautifully in a social media post: "Active listening is the most sophisticated diagnostic <u>and</u> therapeutic tool in healthcare."[12]

We see, time and time again, that being heard in a medical appointment contributes significantly to a positive healthcare experience and offers healing in itself. On the flip side, a lack of rapport and not feeling that you were heard

---

[12] *Dionisio, Instagram Post @drkarand: March 13, 2023.*

damages the doctor-patient relationship, including the trust a patient has in their practitioner and the treatment plan itself. This is unfortunate, as a rude or disrespectful doctor could be offering the best treatment option given the scenario, but their bedside manner gets in the way.

I'm not saying that all healthcare practitioners need to exchange pleasantries, sugarcoat important aspects of care, or smile incessantly; however, all practitioners should treat their patients with respect and not gloss over pertinent information. The most evidence-based way to provide medical care is one where information and decision-making are shared.

It's no wonder that so many of us feel misunderstood, unvalued, dismissed, unsupported, and even gaslighted when concerns are not validated. There has been a longstanding tradition of placing the importance and care of some medical conditions and some populations over others. We have accumulated a significant body of research exploring dismissal and gaslighting based on gender, race, sexual orientation, age, body size, and status.

This prevents people from accessing care. It has been documented that menstruators, especially young ones, will often take a DIY (do-it-yourself) approach to care instead of reaching out to a healthcare professional, even in places with public healthcare systems. The same applies to marginalized populations.

Of course, there are many wonderful practitioners out there; however, there are also those who say things like:

"Come back when you're trying to get pregnant. There's nothing I can do for you right now."

"Having no libido seems like a small price to pay for not getting pregnant."

"Come back when you've lost some weight."

"It's all in your head. Your blood work and imaging are all normal."

These are the kinds of things patients tell me that have been said to them. There are also studies and tons of anecdotal evidence online along the same lines.

I want you to remember that gaslighting, shaming, bullying, discrimination, and disrespect have no place in (your) healthcare. The way concepts are explained, information is communicated, and treatment plans are created with patients can make the difference between feeling supported versus dismissed. It can be the difference between people accessing healthcare and not. It can be the difference between going to school or not. It can be the difference between showing up for a loved one or not. It can even be the difference between living or not.

In my private practice, I tend to attract many patients who are dissatisfied with their healthcare experiences elsewhere. This is primarily a product of not having time with their practitioners in our overburdened one-concern-per-visit kind of healthcare system and one that isn't particularly equipped to serve the complex nature of health and chronic disease all too well.

I also see a lot of patients who are fairly satisfied with the care from their other healthcare practitioners but are looking for more: they desire another opinion or prefer a multimodal and integrative approach with someone who will take the time to answer questions, offer perspective, and discuss options. As a naturopathic doctor who sits adjacent to the public system, I *can* offer more time, whereas the universal healthcare system (in Ontario) doesn't allow for that. This is also part of the reason patients look online for answers to the riddle that is their health or chronic illness, thinking that they'll find the root cause of their issue when there's no singular root cause, but a web of interconnected contributors.

Much of my time with patients in discussion, which is necessary for diagnosis, monitoring progress, and communicating treatment. I also spend a fair amount of time weighing options, even helping patients comply with medications prescribed by their primary care doctor because their medical doctor didn't have the time to fully describe the importance of that prescription.

I enjoy what I do–assessment and diagnosis, blood tests, discussing and prescribing treatment options including supplements, herbs, hormone therapy, acupuncture, and other integrative therapies, collaborating with healthcare practitioners, educating and offering resources, validating experiences and symptoms, answering questions and debunking myths, having authentic conversations about life and health, and promoting wellness.

I sometimes don't get it right, but I strive to be the best I can, challenging myself to learn from my mistakes (who hasn't made those?!), from patients, from colleagues, and from the research and guidelines that are continually being updated. There are things I do well, but I cannot do it all. It takes a village.

I want my patients to have a primary care medical doctor. My patients need specialists. My patients also need access to the kind of care I cannot provide, which can be a challenge at times. The model of medical care that has existed based on cause and effect, like in the case of infectious and communicable diseases, does not serve chronic diseases.

Medical practice is always evolving, and it should change over time. I don't practice the way I used to eight years ago. I sometimes cringe when I see some of my old online articles and even treatment plans. Honestly, even in the last few years, as I've written this book, read hundreds of research articles, sifted through my thoughts, and waded alongside patients during the COVID-19 pandemic and global strife, my practice has changed; I have changed. My approach to care is more open-minded, my treatment plans have become more practical, and my judgment on infrastructure and politicians is more critical.

Clinical decision-making is not clear-cut. In fact, it is quite complicated. We don't often see the messy bits or talk about all the layers publicly. And, unfortunately, we live in a time riddled with polarization and unsolicited commentary from total strangers without any idea of the goings-on in someone's life seems to be the norm (... how did we get here?!).

Health isn't black and white; it's constantly changing. Sometimes, we have a clear path. Other times, however, there can be inconclusive data, not enough research, only not-great options to choose from, adverse reactions, and other considerations. This is why we need more robust discussions in the doctor's office and more collaboration among healthcare providers.

Ultimately, you are the most important part of your healthcare. You have the right to ask your doctor questions about your treatment plan and monitoring, your options, and to access healthcare that will not force your hand or make you feel bad about whatever decision you choose to make. This is all part of informed consent in your doctor's office.

BOX 16. What is informed consent?

Informed consent is about you being a part of your healthcare. It's about making educated decisions on how you want to proceed with your care under the guidance of your healthcare team.

This includes what the intervention is, the pros, the cons, what other options exist, and what happens if you do nothing. It includes your capacity to make a decision, and it includes voluntariness, meaning freedom from coercion or other influence.

Informed consent is a concept relevant to diagnosis, treatment options, referrals, conducting blood work, and much more. And, it is not a one-time thing; it is always part of the conversation, and it's an important part of the conversation that facilitates shared medical decision-making.

We have seen time and time again that there is often a misalignment between what medicine sees as important–the condition or disease–versus what is actually important–the patient. This is especially true of menstrual conditions. We have menstruators who are missing out on what they want to or could be doing for anywhere from one day to two weeks per month, which equates to almost two weeks and up to six months per year of their life!

This is even more significant when we consider how lacking our education systems are and how little the average person learns about the body.

There are some pilot projects underway to disseminate simple health facts and evidence-based lifestyle strategies to the public. One of these is an old-school type of letter campaign sharing the natural increase in cardiovascular disease risk in women after menopause, along with simple strategies like reducing red meat and ensuring calcium adequacy. It doesn't have to be complicated. We can even utilize the online world better to deliver facts and create community–a shift that has been happening quite naturally.

Healthcare providers can make it a point to offer resources. I, firsthand, know that this works because I see it in practice. It's also why I wrote this book! Most of life and health happen *outside* of the doctor's office.

I offer patients vetted resources like books, podcasts, PDFs, and websites alongside having discussions about their long-term health all the time. Much of health requires repetition, which means we need regular exposure to pertinent health information.

Everyone deserves the right to respectful healthcare. It has been documented that women historically were willing to undergo extensive medical or surgical treatment despite a lack of visible pathology to have a better quality of life. We deserve better than that.

As the healthcare system (hopefully) and people within it (slowly but surely) evolve, what you, as a patient, can do is advocate for yourself. Ask your doctors for resources and information. If they don't have any for you, suggest that they start putting some together. I've compiled a list of self-advocacy considerations for you in Chapter 31 to make your healthcare appointments more effective.

Your health matters. You matter.

CHAPTER HIGHLIGHTS

- The biggest barrier we face in healthcare is systemic in nature, i.e., the capitalist model that puts profit over people, our planet, and the future.
- There is much room for growth in our understanding and treatment of menstrual and hormonal health conditions. We've come a long way, but need more data that is diverse and inclusive so that more of our menstruating population can be supported better.

- Much of health and life happens outside of your medical appointments, which means that regular exposure to relevant health information is required for you to improve your health.
- Hormonal health concerns are sometimes normalized and dismissed, which can pose challenges to accessing care. Healthcare providers need to do more listening, validating, learning, and referring for people with periods to not miss out on life. The healthcare system needs to support practitioners in better ways as healthcare personnel have a large burden of responsibility.
- Making decisions about health is easy at times while being difficult and confusing at other times. Health and life are dynamic, which is why we need more discussion and informed consent in medical appointments. Informed consent, essentially, is about you being a part of your healthcare and decision-making process.

CHAPTER 24

The P Word

As I mentioned in the Introduction, my first period caught me by utter surprise. I didn't know it was a thing. I had never stumbled across my mum's stash of pads. I thought that my friends would check the backs of their school uniforms for gum or dust from our classroom benches, and I didn't realize that those advertisements on TV with absorbent blue liquid had anything to do with me.

Frankly, I didn't know the Urdu word for menstruation until I started writing this book, even though I've grown up speaking Urdu. The two words for periods in Urdu are "haiz" and "mahwari." I'd heard the word "haiz" a few times growing up, but if you had asked me what it meant, I wouldn't have been able to tell you. Mahwari, similar to the root of the word menses from Latin ("mensis"), refers to month or monthly.

We used to call our periods, well, periods. But it was never in our regular voice. "I'm on my period." The words would be said quietly, barely above a whisper, and only after a glance around the room to check for stray men hanging about.

Once I moved back to Canada from Karachi, I started hearing a few different euphemisms for menstruation.

From Aunt Flow and Lady Time to The Curse and Shark Week, there is no shortage of terms that describe the state of menstruation in languages across the world. The International Women's Health Coalition and the Clue app, both based out of the US, conducted a survey where 90,000 people from across the globe responded with thousands of euphemisms for periods in different languages–some of them are quite interesting if you want to go take a look![13]

Words have power. According to Alma Gottlieb, a menstrual researcher and author: "Words tell a story. So do efforts to avoid words. The discursive act of substituting euphemism for certain words brings us straight to the territory of taboo."[14]

For most of us, the menstrual taboo is deeply ingrained. When something is so deep-seated that it's second nature, you have to actively work on uprooting and rejecting it.

Periods have been shrouded in a *culture of concealment*, a term explaining how the stigma and taboos around menstruation and period products shape menstrual experiences.[15] The invisibility and stigmatization of periods, most of which have been embedded in us through our environment, impact our menstrual experiences.

The taboos and narratives that have existed around menstruation are complex and culturally diverse. It's easy to blame everything on the male-female hierarchy that has existed (and I'm always happy to go on a tirade about the patriarchy); however, there is a bit of nuance here.

Individuals and communities perceive and experience menstruation in strikingly different ways based on religion, politics, demographics, economics, culture, language, geography, and familial upbringing.

The book *Blood Magic* by Thomas Buckley and Alma Gottlieb offers an anthropological view on the views on menstruation in cultures across the world if you're interested in this topic. In some cultures, menstruation is celebrated as a time to come into your own power, tap into your (body's) wisdom, and connect with fellow menstruators; in others, there is an indifference to

---

[13] *International Women's Health Coalition and Clue App, Top euphemisms for "period" by language, 2016*

[14] *Gottlieb A. Menstrual Taboos: Moving Beyond the Curse, 2020*

[15] *Houppert, The Curse, 1999*

menstruation. And then, there are ones where there are rules around what can or cannot be done during menstruation, sometimes with a negative connotation.

BOX 17. Words have power and they can be confusing

Growing up, I associated periods as a time of being unclean, not being able to do certain things, and hiding what was happening.

I come from a Muslim family. In Islam, similar to other cultures and religions, there are rules around menstruation. I don't remember my mum teaching me about the purpose or science of a period (mainly because she didn't learn that either). I do remember her telling me that I can't fast, pray, or touch the Quran, and that there's a special cleansing shower that is supposed to be carried out after one's period is done. These themes of uncleanliness and exclusion are present in many parts of the world.

To clarify slightly (without getting into a topic I don't know too much about), menstruation is a time of being *najis* or *najasa* in Islam (these are Arabic terms), which can be loosely translated as being ritually impure. This concept of being ritually impure exists in other religions as well.

In Urdu, however, *najis* means contaminated or polluted, while *najasat* means unclean and dirty. As you can imagine, I, like many others, took all of these terms and meanings as one and thought period blood was impure and unclean ... that I was impure and unclean when bleeding, not just physically but spiritually as well. The term "feminine sanitary products" didn't help. It has taken me many years to extricate the deep-seated connection between period blood and being impure and contaminated.

The behaviours and narratives that we learn in our formative years teach us that periods are not meant to be public. Studies conducted in teenagers demonstrate that the messaging that is received is to live as if you are not menstruating. I can totally relate to that! When not fasting during the month of *Ramzan* (or *Ramadan*), my cousins and I would make up random excuses instead of saying we were on our period. This was learned behaviour–no one told me that I needed to do that. I just knew.

Right from the get-go, we start using euphemisms, averting our gazes when buying period products, hiding menstrual products up our sleeves, tearing off product wrappers with the stealth of a ninja, normalizing period symptoms that affect our quality of life, whispering in our friend's ear to check our butts for leakage, and being ashamed of leakage and periods in general.

Let's take a detour and discuss shame for a minute. Periods, a completely natural bodily process, can contribute to feelings of embarrassment and shame. As discussed, this stems from a lot of places, including a culture of concealment, being mocked by peers, ridiculed by society, the patriarchy, the use of menstrual and anatomical terms as pejoratives, and vulgarisms used in place of reproductive anatomical terminology. If the mind is constantly receiving these signals and periods are a monthly experience, then shame and negativity are also monthly experiences, even if only subconsciously.

Brené Brown, a researcher and author, says that shame is an intensely painful experience of believing we are flawed and thus unworthy of acceptance and belonging, and that shame resilience can be fostered by sharing our stories–the good, the bad, the ugly, and the beautiful–with one another.[16] That's why using your voice and sharing your story is so vital. When you share, you start to break down the walls of shame; when you share, you chip away at the systems that rely on us feeling like we have to hide parts of us away.

The words you speak and the thoughts you think have an impact on your experience of your menstrual cycle. The truth is that periods are a fact of life. Through awareness, education, advocacy, and communication, we can start to shift the narrative for ourselves and others. I know this is already happening in many parts of the world, especially online, which is great.

---

[16] *Brown, Daring Greatly, 2018*

Some adolescents go into puberty with excitement and anticipation, some with an indifferent attitude, some with all the feels, and others, like me, go in utterly clueless.

Global research demonstrates that many adolescents go into puberty with a negatively skewed view of menstruation–from anxiety and fear to sadness and shame. A lack of education and a lack of resources are a big part of these feelings. And if you don't identify as female, puberty can be a time of significant discomfort and dysphoria.

We have studies and real-life evidence showing that when people with periods understand their cycles better and have access to resources, it improves their quality of life. The same goes for neutral period education that is provided to *all* adolescents regardless of whether they menstruate or not. Some studies show that empathy, support, and understanding from partners improve well-being by reducing self-pathologizing in menstruating women in heterosexual relationships. When you start to shift the narrative around periods to more neutral and positive dialogue, there's an improvement in quality of life.

Talking about periods to other menstruators also provides something to bond over. Many people connect over premenstrual symptoms, discomfort, and perimenopausal changes. The online world is rife with communities dedicated to period problems and the sharing of resources, which is incredible. It helps so many people feel validated and less lonely. It also advances advocacy and awareness. This has been the beauty of the online world–promoting awareness and mobilizing campaigns. That said, I hope we continue to work at using more positive and neutral language while bonding over, but no longer normalizing, common symptoms that get in the way of life.

Shifting the conversation to a more wholesome dialogue that encompasses positive and neutral messaging will hopefully allow us to break away from the patriarchal and medicalized view of menstruation as problematic and something to hide.

So, talk more about periods and get curious about the kinds of thoughts and words you associate with your menstrual cycles. Can you reframe them to something more neutral or positive? Maybe something a bit more light-hearted? Do you need some help with that? Can you talk to someone in your life about this? Can you grab a resource that helps expand your horizons a bit?

2015 was deemed the Year of the Period by National Public Radio (NPR) in the US. That was the year #periodsarenotaninsult started trending. Since then, we've seen a lot more coverage on periods–youth activism in schools, changes in taxation on period products (which used to be, and in many places still are, classified as non-essential items), free period products in schools, menstrual leave, medical research on menstruation and in menstruating populations, period product drives, movies and books, Olympic athletes and celebrities talking about their periods and hormone-based diagnoses, and so much more.

As a society, what we need to do is improve period education at different stages of our menstruating years. We need to put matter-of-fact age-appropriate menstrual education in curricula everywhere so that menstruators and non-menstruators alike can have an understanding of this basic biology and can continue to develop that over time.

Medical offices can actively offer vetted resources for menstruators to continue learning about their bodies throughout their lives–perimenopause and menopause take so many people by surprise. There is only so much that can happen in a doctor's visit. Credible information offers patients an opportunity to reflect on their health, seek care when needed, have agency over their health decisions, and not be held back by menstrual problems.

This is how we can move forward and change lives for the better. We have research that explores this. I have seen all of this corroborated in my life and my practice. So, let's bleed out loud. Go ahead, and live like you *are* menstruating.

CHAPTER HIGHLIGHTS

- Periods have historically been shrouded in a culture of concealment. They have been hidden from view and discussed through euphemisms in hushed voices. The tide is changing with more coverage in the news, through books and research, free period product legislation, menstrual leave policies, and more open discussion online. We've got to keep moving forward!

- The taboos and narratives that have existed around menstruation are complex and culturally diverse. Part of it is the patriarchy, but the perception and experience of menstruation are different across the globe based on religion, politics, demographics, economics, culture, language, and familial upbringing.
- Research shows that when menstruators understand what is happening in their bodies and during a menstrual cycle through education and age-appropriate resources, it improves their quality of life. So, in addition to talking more openly about hormonal and menstrual health, we need to work on high-quality and age-appropriate resources for everyone so that this very natural part of life does not get in the way of living life fully.

CHAPTER 25
Period Products

A book on periods wouldn't be complete without a chapter on period products. I, like most healthcare professionals, did not learn about menstrual products or free bleeding in school. I learned a lot as I researched products (but didn't find too much scientific data on the subject), explored online and brick-and-mortar stores, explored forums and blogs, and talked to people about them.

I'm going to focus on commercially available products here. There is a huge variability in home-made products with potential hygiene and safety issues and differing degrees of absorbency.

Period Product Types

Period products range from internal ones like tampons, cups, and discs to external ones like pads, liners, and underwear. Free bleeding isn't a particularly common practice as far as I know.

The choice of which product to use is an individual one. Many people employ a few different products, sometimes simultaneously, over their period. If an internal product is causing pain or having difficulty going in, it's best to talk to a healthcare provider to ensure that there aren't any structural contributors to this challenge.

Quantity of flow, length of period, comfort, availability, cost, skin sensitivity, ability or disability, geographic origin, socioeconomic factors, education, culture, and religion are all factors when choosing products.

Tampons were a no-no in my household. In fact, in countries like Pakistan, there continues to be a low uptake of tampons due to "virginity" considerations. Another reason for a lack of products that are inserted into the vaginal tract is the negative connotations associated with periods, menstrual flow, and one's anatomy. I actually didn't know that anything beyond pads existed when I lived in Karachi.

Pads and liners typically have three layers: a top sheet that is semi-permeable to allow for dryness, an absorbent core, and an impermeable back sheet that usually has adhesive (and even wings to keep the pad in place). Cellulose is the most commonly used material, although there are other products on the market like those with hypoallergenic and antibacterial top layers made of corn as well as polymeric foam core layers. Pads are available in many varieties such as those with fragrances (which may cause irritation), and emollients (which are compounds that soothe the skin). Most people tend to change pads before they are fully saturated, and the amount of pads used is not necessarily correlated with the amount of blood loss. This is one of the factors that makes it difficult to gauge how much blood is lost during a period.

Tampons have been made from different materials around the world dating hundreds of years ago: paper, soft papyrus, native plants, wood, and more. The tampons of today are typically made of rayon, cotton, or a mix of the two, and expand vertically when they absorb fluid. They are inserted into the vaginal tract using a plastic or cardboard applicator, or they can be inserted using clean fingers. They have a string to facilitate removal. There have been incidences of people forgetting to remove tampons or being unable to do so. These situations are uncommon but happen more frequently than you'd think. They tend to occur if a tampon is inserted while another one is still in or during penetrative intercourse with an inserted tampon. A healthcare professional should be able to help you out with this. The most common sign of a retained tampon is a foul smell; an infection can ensue if there is a retained tampon, so it's best to keep track of your tampons.

BOX 18. What about Toxic Shock Syndrome?

Toxic Shock Syndrome (TSS) is a fairly uncommon health concern that can occur in males and females. TSS is caused by TSS toxin 1-producing *Staphylococcus aureus* bacteria as well as some *Streptococcus species*. TSS can be non-menstrual–typically related to soft tissue infection or post-influenza–or it can be menstrual (mTSS), which happens immediately before, after, or during menstruation. *Staph aureus* is the bacterium that's the culprit behind mTSS.

Data from the US estimates that the incidence of mTSS is 0.5 to 1 per 100,000 people, with a maximum incidence of 10 per 100,000 people.[17] It usually manifests as fever, low blood pressure, rash, and dysfunction of various organs like the gut, kidney, liver, and brain. It is treated with antibiotics.

Some people have immunity to developing mTSS through antibodies and a lack of a response to the toxin, while others are slightly more prone to it. The introduction of oxygen through internal period products like tampons, cups, and sponges is a potential link to developing mTSS as is the slight increase in pH during menstruation.

It was a more popular topic in the 1980s as the Rely tampon was associated with higher rates of mTSS. Rely was pulled off the market consequently as were polyacrylate, carboxymethyl cellulose, and polyester foam tampon products due to investigations and research conducted by the Centers for Disease Control and Prevention (CDC). The guidelines to have warnings and instructions of use (including recommended duration of use) on tampon boxes were also introduced at this time.

Cups and discs have been gaining popularity over the last five years. These products collect blood instead of absorbing it. Menstrual cups sit in the vaginal canal via suction, while menstrual discs sit at the base of the cervix in the area called the vaginal fornix, where the cervix meets the vaginal canal. Research and

[17] *DeVries et. al, PLoS One, 2000-2006.*

anecdotal evidence from patients indicate that they are comfortable and effective.

Reusable cups and discs are typically made of medical-grade silicone, thermoplastic elastomer (TPE), or rubber. Disposable ones are made of medical-grade polymers, with innovations trying to use plant-based and carbon-capturing polymers (instead of carbon-emitting ones). If you use a reusable disc or cup, it's best to have at least two so that you can clean and sanitize appropriately between uses. Washing and rinsing with a mild oil-free cleanser is recommended on the regular when using it during a period while sterilizing in boiling water is recommended before and after a period. Each reusable cup and disc comes with care and cleaning instructions and it's best practice to follow the ones that came with yours. Cups and discs need to be folded for insertion: C fold, punch down-fold, seven-fold, S fold, and triangle fold are some of the options for insertion, with instructions available with the product that you get. There are even menstrual applicators where the product is inserted into the applicator and released once the applicator is inside the vaginal canal. Cup and disc removal can be a little messy (or a lot!) the first few times, so it's best to use a private bathroom where you have some time to do so. These products can be used for up to twelve hours, which a lot of menstruators like.

Period underwear has also become quite popular, and I personally really like these for my low-flow days. They are typically made with cotton or nylon and are designed like pads with a moisture-wicking top layer, absorbent core layer, and leak-resistant bottom layer. A small 2023 study comparing period products showed that period panties hold the least amount of menstrual blood, with an overestimation of their listed absorbency.

Reusable pads have a similar concept and tend to be made of cotton or bamboo with varying absorbencies, and they tend to have wings with buttons to hold them in place around underwear. Care instructions for both of these are to rinse upon soiling, and then hand washing or washing with delicates in cool water and then hang dry. Cloth period products are typically less irritating to the skin than other products, so they are good options for anyone with skin sensitivities;

however, they're typically more costly than disposable products (and a lot more costly compared to regular underwear).

Period Product Practicalities

The aforementioned products come in different varieties. The standardization of period products based on the amount of fluid absorbed is as follows:

- Junior: less than six grams (around one teaspoon)
- Regular: six to nine grams (around 1.5 to 2 teaspoons)
- Super: nine to twelve grams (around 2 to 2.5 teaspoons)
- Super plus: twelve to fifteen grams (around three teaspoons or one tablespoon)
- Ultra: fifteen to eighteen grams (over one tablespoon)

Period underwear and reusable pads will usually express their absorbency in relation to these standards as well, even though period underwear tends to underperform. Menstrual cups and discs are not well-standardized from company to company; however, they come in different sizes that can hold ten to thirty-eight millilitres of fluid. In the aforementioned study laying out the volume of blood contained, discs were able to hold the most amount of blood. Trying out one or two sizes is fairly common when it comes to cups to find the best fit that is comfortable. Leakage is one of the most pressing concerns for most menstruators, so using an appropriate product that meets your needs is important.

Cost is a factor when it comes to period product selection. There is a huge range in price, as most of you probably know. Liners can average from six to fifty cents each, disposable pads from ten cents to one dollar and fifty cents, reusable pads from two to twenty dollars, tampons from twenty cents to two dollars, reusable cups from twenty to fifty dollars, reusable discs from fifteen to thirty dollars, disposable discs from three to five dollars, and period underwear from fifteen to over fifty dollars (all prices are in Canadian dollars, excluding tax).*[18]

---

[18] *Prices taken from Walmart Canada and Amazon Canada, 2023*

We have statistics that on average, a Canadian menstruator uses approximately 169 disposable products and spends, on average, seventy-five dollars plus tax per year.

This number will change based on your personal needs and preferences. And the household cost, obviously, is way higher when you have multiple menstruators in the same home, have recently given birth, or have heavy or prolonged menstrual bleeding. There is a huge push for government subsidization and universal access to period products as these are essential to us menstruators. More on that in Chapter 26.

If you were to add up the number of days that menstrual blood loss is experienced, the average comes out to be five to seven years of life. Essentially, the average menstruator cumulatively bleeds for an equivalent of six years of life with over 5000 disposable products used.

Traditional single-use pads and tampons end up in waste and cannot really be recycled or reused. If you have the means, a reusable or compostable product would help with waste reduction even if used infrequently or inconsistently. But remember, this isn't the only way you can contribute to reducing your waste, especially when we consider that big corporations are the leaders in waste production and contributions to climate change.

Menstrual product regulation varies across the globe, where some countries have strict criteria to meet as medical devices, while in other countries, general product regulations have to be met.

In Canada, menstrual pads fall under consumer goods while tampons are regulated as medical devices with Health Canada ensuring safety, efficacy, and quality. Packaging must contain specifics about absorbency and the risk of TSS. The recommendations from Health Canada include avoiding tampons if you've ever had TSS, using the lowest absorbency to meet your needs, reading the instructions, using them for menstrual flow only, practicing good hygiene before and after insertion, not using them overnight and changing every four to eight hours, alternating with external protection like pads, and remembering to remove each tampon.[19]

BOX 19. What about dioxins?

Talk of dioxins comes up in period product discussions in the online space. Dioxins are hormone disruptors and carcinogens (cancer-causing compounds). Dioxins in period products used to be a result of the chlorine bleaching process while manufacturing; this has been discontinued. However, since dioxins are so prevalent, trace amounts are present in cotton (even organic cotton) and pulp-based materials like rayon.

So far, studies have shown that these exposure levels are multitudes (like, thousands) of folds lower than dietary exposure to dioxins. I'm no environmental medicine expert, but lowering dietary exposure by decreasing animal fat (where dioxins tend to accumulate) is probably a good place to start to decrease exposure to dioxins as we wait on more data to draw any conclusions about period products.

The safest way forward, based on research and historic use, seems to be to use a commercially available product as the packaging describes: do not use it for more than approximately six to eight hours, make sure you're using an appropriate size for your flow and anatomy and comfort, incorporate a reusable product if and when possible, and use fragrance-free products. There are incidences of allergic contact dermatitis and skin irritation with period products, which primarily occur from products with fragrances (including essential oils). There is no need to perfume the vulva or vagina.

Most people change their products before they are fully saturated. The risk of itching, irritation, and infection increases when there is prolonged moisture in the vulvovaginal area. The number of eight hours historically proposed for tampon replacement wasn't particularly based on research but offered a good benchmark. A recent study from France (done in a small group of patients) showed that leaving a tampon inserted for more than six hours slightly increased

[19] *Government of Canada, Menstrual Tampons, 2015.*

the risk of developing TSS, so perhaps approximately six hours should be the cutoff.[20]

Affordability and accessibility (of products and sanitation) are also factors. A lack of period products means that menstruators miss out on school, experience teasing and shame, and can potentially contract infections from prolonged use of products. It's hard for people who have no accessibility issues to imagine these scenarios, but these situations are still very prevalent in the world at large in the 2020s.

There are wonderful organizations within Canada and across the globe that conduct period drives and are advocating for free access to period products and ensuring that all bathrooms have disposal bins to ensure privacy, especially in educational institutions. Appendix B has a list of a few of these organizations.

CHAPTER HIGHLIGHTS

- Period products range from internal ones like tampons, cups, and discs to external ones like pads, liners, and underwear.
- Quantity of flow, length of period, comfort, availability, cost, skin sensitivity, ability or disability, geographic origin, socioeconomic factors, education, culture, and religion are all factors when choosing products.
- Pads and liners typically have three layers: a top sheet which is semi-permeable to allow for dryness, an absorbent core, and an impermeable back sheet which typically has adhesive. Period underwear usually has three layers like a pad.
- Tampons expand vertically when they absorb fluid. They are inserted into the vaginal tract using a plastic or cardboard applicator, or they can be inserted using clean fingers.
- Pads and tampons are standardized for blood absorbency and range from junior to ultra, holding six to eighteen grams of blood (i.e.,

[20] *Billon et. al, EClinMed, 2020.*

around one tsp to over one tbsp) of blood respectively when fully saturated.

- Cups and discs collect blood instead of absorbing it. Menstrual cups sit in the vaginal canal via suction, while menstrual discs sit at the base of the cervix. Depending on the company and size, they can hold anywhere from ten to forty millilitres of blood.
- The risk of itching, irritation, and infection increases when there is prolonged moisture in the vulvovaginal area. The safest way seems to be to use a commercially available fragrance-free product as the packaging describes: do not use it for more than approximately six to eight hours, make sure you're using an appropriate size for your flow, anatomy and comfort, and incorporate a reusable product if you can.

CHAPTER 26

Menstrual Equity and Activism

Menstrual activism has seen a big boom in the last decade, which is heartening to see. Challenges to policies, taxation, accessibility, discrimination, and the narrative surrounding menstruation are happening worldwide.

Activism and advocacy have taken the form of education reform, period product drives like Tampon Tuesdays, increased menstrual research, involvement of sanitation and hygiene facilities, community-based family planning programs, universal contraception care plans, movies and documentaries, media coverage, policy change, and more.

Shilpa Gupta's art installation using menstrual blood-stained fabric, Kiran Ghandhi's London Marathon free bleed, and Rupi Kaur's "Period." photo series are great examples of personal activism that brought menstruation into the limelight. In 2019, May 28 was proclaimed as Menstrual Hygiene Day to raise awareness and find solutions to support menstruators regarding the accessibility of period products. It's also called Menstrual Health Day, which is definitely a better name.

We are seeing more and more cohesion in the menstrual movement as well. One of the ways this is happening is through language. As discussed in the Introduction to this book, menstrual health was defined and refers to well-being in relation to the menstrual cycle including accessibility, healthcare, respect, and education. Another term is *menstrual equity*, which is all about equitable access to education, care, products, and resources.[21] A lack of infrastructure–financial, cultural, social, political, and geographical–poses the biggest barrier to menstrual equity.

In the global public health field, menstrual equity has taken the form of Menstrual Health and Hygiene Management (MHHM) as well as the provision of sexual and reproductive health (SRH) programming. This refers to overseeing the factors that impact menstrual experiences such as sociopolitical, psychosocial, and environmental aspects. It includes access to period education and period products and addressing the stigmata around menstruation.

It became part of the public health agenda to ensure that young girls would not be discouraged from attending school due to fear and shame. Instead, a sense of confidence in their abilities to manage menstruation and have agency over their sexual and reproductive health would be fostered. It is an area where many strides have been made, but a variety of obstacles still stand in the way, especially but not exclusively, in low and middle-income countries.

According to an upcoming report by Moon Time Connections, a Canadian Indigenous-led period equity group, 70 percent of Indigenous respondents from remote communities in the North have issues accessing products.[22]

According to a 2018 report from Plan International Canada of 2000 Canadian women, 30 percent of the women surveyed said they've struggled to afford menstrual products, and 83 percent say they've missed school, work, or social activities due to their period.[23]

Factors to consider here include education, affordability of and access to period products, menstrual concern management, garbage and disposal facilities, as well as access to clean water, sanitation, and hygiene (WASH) services.

---

[21] *Weiss-Wolf, Periods Gone Public, 2017*

[22] *Moon Time Connections, Instagram Post @moontimeconnections, Dec 18, 2023*

[23] *Plan International Canada, A Canadian Gender Study - Period Stigma Report, 2018*

Hygiene and facilities (disposal, cleaning, running water) are still a big issue in many parts of the world. Many of us take for granted having bathrooms and period products of all kinds and the privacy to take care of ourselves. We can go through our periods with dignity. In low-income settings, WASH services, reusable product hygiene, frequency of absorbent product change, availability of soap and water as well as privacy are all barriers to managing periods with dignity and without fear. For others, however, washrooms can be a place of confusion and anxiety, especially transgender and gender diverse folk assigned female at birth as well as people with physical ailments or disabilities.

We've already discussed how menstrual education improves the quality of life of menstruators. Pilot studies have shown that in areas where this education was previously missing and resource availability is low, school enrolment and attendance improve when students are taught about these topics. Social support from peers and teachers seems to be an important aspect of improving the lives and participation of adolescents in school.

Community-based and peer-led hubs seem to be the most feasible and acceptable way of doing this according to studies conducted in various parts of the world like India, Nepal, Uganda, and Zambia. Additional factors included household support and financial literacy. Ensuring that all of this is done in a culturally appropriate way where community members are leaders in the development and delivery of material and infrastructure is, of course, key.

Policymakers have been making history in the last decade all across the world. Many African and Asian countries have broken the silence on menstruation, recognizing the needs of people with periods and introducing policies and institutional changes to health, budgets, education, government guidelines, and a variety of projects. Taxation on period products called the *tampon tax* is slowly being abolished across the world as well. Kenya was the first country to get rid of this antiquated and discriminatory tax in 2004 and other countries like Australia, Uganda, India, Canada, Nicaragua, Malaysia, Lebanon, and the UK slowly followed suit.

In 2020, Scotland became the first ever nation to make period products free and universally accessible for anyone who needs them, placing the onus on local authorities. That's what we need: a national plan that can be shared with local governments to implement in a way that is feasible for them. It also allows for innovation in the delivery of plans. We can draw from the strengths of the different methods and adopt ones that work. Other countries like Botswana, South Africa, France, Australia, and New Zealand have also started providing menstrual products in schools since then.

In Canada, the city of London, Ontario was the first city to offer free menstrual products in facilities run by the city. This was thanks to the work of local advocates including Here for Her, and the Thames Valley District School Board who took the initiative to get things started in local schools.

Since then, we've seen more and more youth-led, community, provincial, and national organizations follow suit to advocate for the same. 2022 brought with it a draft proposal under Part II of the Canada Labour Code to provide menstrual products for workers who need them in the public and private sectors that are regulated federally, which came into effect as of December 2023. Also in 2023, the federal government gave Food Banks Canada 17.9 million Canadian dollars to run a national pilot program to address the needs of Canadians with period product needs. [24] [25]

In light of recent events, where those assigned female at birth are losing access to education and the right to bodily autonomy, we need to rally for each other, especially for those impacted the most by structural inequities.

We've been seeing a rising tide of voices against oppression, which is promising. Simultaneously, we're seeing blatant examples of white supremacy, ethnic cleansing, and hate crimes.

To move the needle forward, we need more menstruators in varied positions in the world–in the home, in the community, in schools, in governments, in

---

[24] *Canada Labour Code, Regulations Amending Certain Regulations Made Under the Canada Labour Code (Menstrual Products), 2022.*

[25] *Liberal Party of Canada, Free Menstrual Products, 2022.*

politics, in science, in tech, in finance, in the media, and in business. We need those who come from a place of love, abundance, and respect.

Marginalized populations always suffer the brunt of policies that are made without them in mind. This is true for health in general and challenges around menstrual equity are no different. These issues affect those with low incomes, the homeless, the displaced, and the incarcerated more significantly than other populations. As the world's richest get richer and have more power and influence, as our governments and policy-makers remain corrupt, and as climate change impacts the exploited the most, we'll see some of these disparities widen unless we address all of these structures that contribute to inequity.

I've written most of this book amid a pandemic whilst discrimination, oppression, genocide, reckless politics, profiteering, and war on basic human rights are happening right in the open. It is the oppressed who usually carry the burdens of their own oppression as well as those of others; they also are the ones who fight for collective liberation.

We need to take action and do better. We need you. We need you to speak up, advocate, and create change for yourself and others. For that to happen, you need to have energy, hope, joy, community, and health. And, that's what the final part of this book is about–how you can move toward more peaceful periods.

CHAPTER HIGHLIGHTS

- Challenges to policies, taxation, accessibility, discrimination, and the narrative surrounding menstruation are happening worldwide. Activism and support have taken the form of education reform, period product drives like Tampon Tuesdays, menstrual research, involvement of sanitation and hygiene facilities, community-based family planning programs, movies and documentaries, media coverage, policy change, and more. I encourage you to get involved as much as you can.

- We need more menstruators in varied positions in the world – in the home, in the community, in schools, in governments, in policy, in science, in tech, in finance, in the media, and in business. This is the way we'll continue to move toward more menstrual equity and better health and healthcare for all.

PART SIX

PAVING THE PATH TO PEACEFUL PERIODS

CHAPTER 27

Bring it Back to the Basics

Now that we've gone through the ins and outs of the menstrual cycle while holding that in the context of the world at large, you might be wondering what you can *actually* do to support your cycles on the regular. These final few chapters will offer considerations for menstrual wellness and overall health.

As you know, menstrual cycles involve a complex and fascinating dance involving players from all across the body. Support for these systems, however, does not need to be super complicated. There are many simple things that we often overlook because the trendy gimmicks and clickbait headlines–not to mention paid ads and propaganda–steal all the attention.

Reducing the risk of chronic illness and long-term health lies in the seemingly mundane things: in the routine of daily life. For many of us, that can include exercise, supplements, meds, traditional medicines, physical modalities, mental-emotional therapies, hobbies, spiritual practices, community, social support, and the like.

My colleague, Dr. Martha Sharpe, ND, says it well: "Prevention isn't about a guaranteed outcome ... we're talking about reducing the load in the bucket..."[26] Adding in health-promoting behaviours is about improving your quality of life now and for later.

---

[26] *Sharpe, Instagram Post @marthasharpe_nd: May 22, 2024.*

I know that many of you are looking for this—to be an active participant in your health and healthcare—whether you have a diagnosis of some kind or not.

Unfortunately, our lack of education on lifestyle variables and existing gaps in the healthcare system doesn't necessarily foster these habits in a practical and age-appropriate way. This can increase the disconnect between how we feel, how we actively participate, and how we want to live. This leads to people looking outward for more, for better, for something else ...

Health in the Age of Dr. Google and TikTok

Having a plethora of information right at our fingertips is wonderful in countless ways. The Information Age has offered community, hope, resources, answers, and direction to so many of us who have felt unsupported, unvalidated, and alone. It has facilitated better health discussions and assessments, even helping patients (like me!) in their journey to get a diagnosis.

There is a lot of robust, nuanced, and accurate information out there. There's also *a lot* of misinformation. Some of this is well-intentioned and innocuous; however, it isn't always harmless.

The amount of orthorexia and disordered eating being promoted online in the name of so-called clean eating is deeply problematic. The same goes for the use of undiluted volatile essential oils, laxative-containing detoxes, and unvetted high-dose nutraceuticals. The internet is rife with scams, multilevel marketing schemes, greenwashing, and the commodification of everything wellness.

If you've found yourself purchasing products or services promising you an outcome that ended up being a waste of time or money, you're not alone (I'm right there with you!). Marketing experts are adept at preying on emotion. Most of us make decisions based on how we're feeling.

There's nothing wrong with you if you're seeking clarity, especially if your healthcare is lacking. Most of us are looking for ways to feel better and to understand what's going on. It has never been easier to get caught up in all the new tech, fads, quick fixes, and before-and-afters that people use to sell things. I totally get it!

Part of my job now includes answering questions my patients have from their time on Instagram and TikTok. I love that my patients are using discernment and bringing in their questions instead of following the regimens and recommendations from people of influence, aka influencers—influencers who curate what they're sharing, who are often paid to share, and who are also simply people with their own unique sets of backgrounds and experiences.

As you navigate the World Wide Web, please exercise caution and put on your critical thinking cap. Take a pause, talk to someone or seek out some different perspectives. In the age of TLDR (too long, didn't read) content, nuance gets lost often. This is also a way to notice any emotions coming up for you.

There are dynamic layers to health. What worked for Mandy or Layla might not work for you for all of the reasons we've discussed throughout this book.

A DIY approach can be, and has been, helpful for a lot of people. On the flip side, it has also delayed diagnoses and medical care for others, even interacting with medications and creating problems that didn't exist before.

We're in the era of apps and biohacking anything and everything from self-care and exercise to morning routines and nutrition. Life has become pretty complicated with different platforms vying for your time and attention. Apps are there to make life easier; if that's not what they're doing, ditch them.

Claim your time, energy, and attention back. Place it on yourself, your dreams, your passions, and your loved ones.

BOX 20. The balancing act of hormonal imbalance

The terms "hormonal imbalance" and "balancing hormones" are often thrown around loosely in the wellness space to describe a lot of different things.

Much of this is well-intentioned to offer language, insight, and resources for common hormonal disturbances that go underdiagnosed and under-supported or even normalized while planting seeds to seek out care. This is great, and I love the internet and social media for it!

However, general labels like these can also be used to sell products, pricey testing, and services to consumers, sometimes needlessly. Not just that, some of this rhetoric has undeservedly tarnished the reputation of vital hormones like estrogen and cortisol while creating confusion and this (erroneous) idea that all hormonal fluctuations are something to fix (hint: they're not!).

Common symptoms that usually fall under "hormone imbalance" can include fatigue, bloating, water retention, hot flashes, acne, headaches, irregular cycles, heavy bleeding, mid-cycle spotting, pain, and cravings. As discussed in previous chapters, some of these might fall under normal responses to hormonal variations depending on frequency and severity. Some, on the other hand, might be quite problematic, disruptive, and indicative of something deeper going on.

The term "hormone imbalance" can likely be defined more specifically. For example, heavy periods and chronic pelvic pain could be endometriosis, cystic acne and irregular cycles could indicate PCOS, and hot flashes with menstrual irregularities and vaginal dryness in your 30s could indicate primary ovarian insufficiency (POI). In cases like these, we want to make sure we're not delaying diagnosis or treatment (while also not spending our hard-earned money and resources on unnecessary testing, products, or services).

This is not a jab because I know so many people have learned and benefitted from online products and services (I'm one of these people!). This is more of a "buyer, beware" call to take a pause when you read or buy into something or someone, especially in today's world of instant gratification, sensationalized headlines, and extreme philosophies.

There's a balance to be found somewhere, I think!

Another area where you likely need to be a bit more analytical when using anecdotal evidence, trying something out, or even following medical advice is to actually understand the nature of what you're doing, how long it'll take, and what you're measuring as progress.

Let's say someone starts taking a supplement for missing periods due to PCOS. They take the appropriate dose but don't know that it'll take two to six months for a change to be seen. Now, many people start getting frustrated after a couple of months of taking something without any results. So, this person stops their supplement after two months and deems it futile. Instead, they start drinking a cup of a tea targeted to PCOS patients daily, and they get their period a week later.

It's easy to understand why this person might think that the tea is what caused their period to come. However, what they don't know is that the two months on that supplement is probably what made that happen. We sometimes give credit where it's not really due!

We also tend to make changes in bulk, so it can get a bit murky when figuring out what works or not. That's not to say that you have to try one thing at a time. It's to say that sometimes our interpretations could be off-base.

You Are Not to Blame

As you navigate your menstrual health, please remember that you are not to blame for all of your health problems.

In the health and wellness space as well as in the wild webs of the internet, sometimes well-intentioned messaging can make you feel like the onus is all on you for what you're going through.

This oftentimes leads to a downward spiral of all the "could'ves" and "should'ves" as well.

Just because there are links between certain lifestyle and nutrition habits and your health, these do not mean that you caused all of your health problems.

Can some of your lifestyle be contributing to your health issues? Sure. Did your past have a role in getting you to where you are? Yes. Do you have to be the one who cares the most about your health? Yup. Can working on modifiable lifestyle factors improve your health? Of course.

While these things are true, so are others, like the fact that your genetics, family of origin, childhood, financial standing, capacity, geography, and other determinants of health are contributing factors to your health, well-being, and risk of developing certain conditions.

You have agency over a lot of your choices and your lifestyle depending on who you are and what privileges you hold. However, we do not all have the same twenty-four hours in the day.

A single mother of a teenage son who works full-time from home living in an urban setting with the disposable income to get groceries delivered and hire a personal trainer has a very different reality than a single mother of a teenage son who works full-time an hour's commute away from her home in a rural setting and lives pay cheque to pay cheque, often taking on odd jobs to make ends meet.

There's a lot to unpack in these two examples including the backstory that I didn't include and the interconnected nature of the social determinants of health that we covered in Chapter 22.

If we were to give these two women the recommendation to add in thirty minutes of exercise per day, it's not a level playing field. It's not a lack of want or understanding of the benefits of exercise that the second person might struggle more. It's also not because she's "making excuses", which is a phrase that often comes up with respect to implementing lifestyle modifications.

We do not all have the same access to resources to implement and maintain change. I think we could all come up with an infinite amount of permutations accounting for the differences between people.

It's because these complexities of health and life are lost on some that we see so much fearmongering and finger-pointing these days. It leads to guilt, self-blame, and confusion, which needs to be dispelled. I hope you have compassion for yourself and others.

You can't change the past, but you can influence the future, majorly in some cases. Wherever you've landed, that's where you are. This is something I say to patients (and myself!) because it's the reality we're working with. We do the best we can with what we know and what we have.

Moving Forward with Patience and Practice

My mantra in my clinical practice revolves around net forward movement, i.e., progress over perfection. Slow and steady wins the race of health promotion and risk reduction.

Creating habits and routines is hard! Many of us forget that getting better at something requires us to fail at it ... a lot!

Moreover, there's this common but bizarre phenomenon where we expect ourselves to be good at anything and everything we try even if we've never done it before! I'm guilty as charged.

Of course, we also expect to do it on our own without missing a beat while fulfilling all the other demands in life. Sounds totally realistic, right?!

Sarcasm aside, this kind of thinking messes with our minds. It sets us up for major failure. We often turn our unsuccessful attempts, aka the learning process, into: "I'm a failure".

Learning is not a linear process. Taking imperfect action regularly is how we actually get better at making changes and cultivating habits. Having successes and setbacks is part of the journey.

Without taking any steps, you'll end up staying where you are. This is totally fine if you're happy with where you're at! If you're looking to create change and move forward, however, then you've gotta roll with the punches and put one foot in front of the other.

Make specific goals so that you have a target to move toward or away from. Science shows that behaviour change comes from wanting to move toward something positive or away from something negative. Oftentimes, we need both.

You can anchor your exercise goal to something positive, like lifting your kid or grandkid on your shoulders, while trying to minimize the back pain you get from your sedentary work. Tethering your health habits to outcomes you want or want to avoid is something you can start to think about.

Stacking your habits is a powerful way to create new habits. This is essentially tying a new habit to one you already have. For example, if you want to add a breathing exercise to start your day, you can tack that onto your morning skincare routine or your morning beverage of choice. If you want to start stretching your neck because it helps with shoulder and neck tension mid-day, you can tether that to your lunch, doing your stretches before or after you eat.

I'm sure there are a lot of things you think about changing in your day-to-day. I encourage you to note these down somewhere and get clear on what it is that you want to start with. Just because you can do everything doesn't mean you must or should. Then, you'll likely need to figure out how long it'll take to see change, what kind of effort it requires, what progress looks like, and how you actually get to that place.

This is where you might need to factor in the high up-front "cost of acquisition": time, effort, money, accountability, letting some things go, asking for help, etc.

To get better at something, you likely need to put in more of yourself at the beginning until it becomes easier or more habitual. It probably also needs to go into your schedule.

Recruiting help can lead to faster change. In the exercise example, this might look like hiring a personal trainer, even if it's only for a little while to show you the ropes and help set your expectations.

I remember when I was learning to play my saxophone. It was hard. I was learning to read music and play an instrument at the same time. Having a dedicated teacher, daily practice, classmates who were on the same journey, at-home resources, and a goal to be in the school band is how I got really good really fast.

I'm not as good as I once was because I simply don't play as much anymore. Whenever I do start playing more, I get better. As in, practice and maintenance are important.

Each one of us lives a different life where our unique set of genetics, experiences, values, education, culture, and everything else converges. Just like there's no singular "right" way to live life, there's no absolute way to take care of your mind, body, and soul either. Health, like life, is dynamic and ever-evolving.

It can be overwhelming to look at *all* of the things you could possibly do. I don't think you need an overhaul or a fancy strategy. Simple is not sexy, but simple is sustainable, and simple tends to work.

For growth, a mix of low-effort, medium-effort, and high-effort activities and accomplishments tends to yield results over time. This approach allows us to build confidence and trust in our ability to try new things, to progress and get better, to fail and live to see another day regardless of performance, and to achieve what we set our minds to (and beyond!).

We are often capable of more than what our brains are telling us. Sometimes, there can be a lot of resistance even when you know that something is good for you. This is not uncommon, and resistance isn't always bad.

I like exercising and rationally know the benefits, but sometimes I *really* don't want to go to the gym. To overcome the resistance, you sort of have to move through it–practice makes it a bit easier. My brain tells me all the things I could do with my time instead of working out, like lying in bed and cuddling with the dogs, doing some work, reading a book, cleaning the washroom, cooking a meal, calling my mum, admin work ... honestly, *anything* else! What gets me into the gym (most of the time, anyway) is how I feel during and after I work out, the sense of accomplishment, the progress I'm making, and the commitment to my present and future self.

Your brain will sometimes get in the way, over-complicating matters and telling you things that aren't necessarily true. Again, practice (and maybe some accountability or help) allows you to do a little bit of what you want to accomplish over time even on the days when your brain's like: do I even care about this? I don't want to do this (anymore)! or it's taking too long. This was the story of my life writing this book!

You can do hard things. The fact is: you've probably already done a lot of hard things in your life. As mentioned already in this chapter, you can't get better at all the things all at once, but you can choose where you set your focus.

The next chapters are comprehensive, but by no means are they exhaustive. As you go through them, I encourage you to identify the habits that are serving you well (and celebrate those!), areas where there's room for improvement, and aspirational practices that you'd like to add to your life. There is deliberate room for nuance because we're all different with different needs, different paths, and different definitions of health and wellness.

You know yourself and your body the best. You probably already know what kinds of things you need to work on (and a few that you could do without!). If only knowing were the same as doing, eh?!

Life is short, but it's also long.

Be intentional in where you put your attention and effort. You can't expect different outcomes when you keep doing things the same way. You will need to create time and a strategy for the things you want more of.

Ask yourself whether or not your goal makes sense right now. If you're in a season of life during which reaching a goal is highly unrealistic because you don't have money, time, help, or mental and emotional capacity, then staying the course or picking a different target is totally okay. You can come back to the original goal when you can put in the effort required to meet the change desired.

Start where you are and move from there. Take a pause whenever you want and feel free to cruise along the way. There are times to push forward, times to pull back, and times to simply be.

CHAPTER HIGHLIGHTS

- Health is complex and multifactorial. As you navigate your menstrual health, please remember that you are not to blame for all of your health problems. There are some things under your control, while others that are not.

- Each one of us lives a different life where our unique set of genetics, experiences, values, education, culture, and everything else converges.
- Long-term health lies in the routines of daily life. Focus on taking one step at a time, creating net forward movement. By really focusing on one or a few things, the results will also be more streamlined and pronounced.
- Layer in the low-effort habits with more difficult ones. You can do hard things!
- There is a time to blaze ahead and a time to stay the course. See where you're at and take it from there.

CHAPTER 28

Cyclical Living

The real reason why you don't feel the same way every day is that you're a cyclical being–we all are. Hormones ebb and flow. Energy waxes and wanes. We go through seasons in our lives. Cyclicity is the nature of life, the universe, and everything. These are the themes that cyclical living draws on.

It's not a new concept and it goes by many names with different iterations. You might be familiar with the term "cycle syncing" which has become very popular online in the last little while; this is a form of cyclical living as well (that we'll get to later in the chapter).

Cyclical living, to me, is more about the overarching patterns in how you feel, how you connect (with yourself and your surroundings), and how you move through life as a whole. It's about acknowledging that cyclicity is part of life, noticing what's happening over your menstrual cycle, tapping into those natural rhythms, and knowing that one part of a cycle flows into the next (see what I did there?!).

Your hormones are dynamic; they affect your life and your life affects them. The natural changes in hormone levels over the menstrual cycle have an impact on mood, digestion, metabolism, energy, recovery, and other aspects of your being. The extent of these changes differs from person to person and they tend to change over time (as things inevitably do).

Cyclical living is not about changing all your routines every time you enter a new menstrual phase. It's not about fixing what's not broken either.

Understanding our bodies and lives through the lens of cyclicity is an opportunity to shed the narrative that there's something wrong with *us* when we don't feel exactly the same throughout the month. It's about cultivating curiosity and awareness. It's an opportunity to assess (or reassess) your relationship with your body, productivity, rest, and your routines. It's about pacing yourself and living with intention. It's a chance to see what's working and what needs work. It's about supporting your health and yourself now and for later.

The Seasons of the Menstrual Cycle

The parts of the menstrual cycle have been likened to the seasons of the year. In many parts of the world, they are associated with the phases of the lunar cycle. Each season, or phase, brings with it a different vibe. They all are dynamic and interdependent parts of a whole that do not exist in isolation. For ovulation to take place, you need the events of the follicular phase to occur, and a luteal phase cannot exist without ovulation. This truth is inherent to cyclical living.

There are many different ways of understanding and embodying this cyclicity. We can take inspiration from history, different cultures and places, and from nature; animals and plants adapt and do things differently each season.

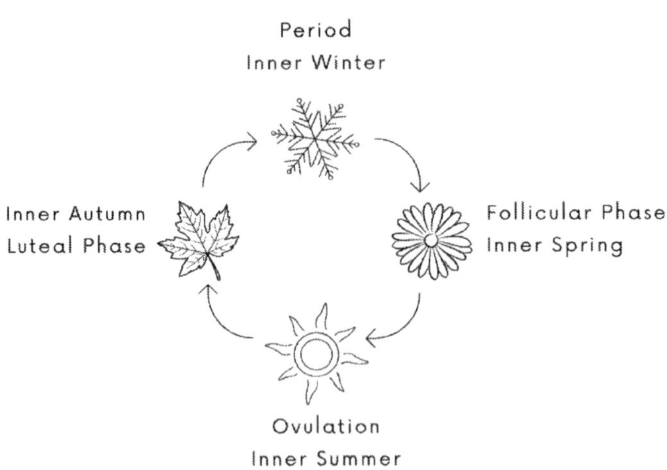

DIAGRAM 10: The *Seasons of the Menstrual Cycle: period (inner winter), follicular phase (inner spring), ovulation (inner summer), and luteal phase (inner autumn).*

I'm going to share an interpretation (as visualized in Diagram 10) that has been informed by a variety of sources; some of what is presented here is adapted from Traditional Chinese Medicine and various Indigenous methods of connecting to one's cycle, some comes from menstrual cycle and hormone research, some from anthropological studies, and some from the lived experiences from menstruators across the globe (personal contacts, books and media, and online forums).

Menstruation: Inner Winter

From a Traditional Chinese Medicine (TCM) perspective, the time when you're bleeding is the most Yin part of the menstrual cycle. Yin pertains to feminine energy that is nurturing.

This is the time when slowing down, rest, reflection, mineral-rich foods, and adequate sleep are most nourishing to the body, mind, and soul (and your Yin itself!). In some cultures, this has been a time to connect not only with yourself but with other menstruators and the earth as well.

It's the winter or new moon of the cycle: a time to slow down, plan and set the tone for how you want the rest of the cycle to look, and build up your stores so that you have the resources for spring, which is a time of growth and change.

Hormonally, this is when estrogen is about to rise and progesterone is low. As a result, you may feel a bit more inclined to take it easy and go with the flow (pun fully intended!). If you have premenstrual syndrome (PMS) or a premenstrual disorder, this season may offer a reprieve from the emotional and physical symptoms of the luteal phase, giving you a chance to recover and get your bearings before moving forward.

Just like your endometrial lining is being shed and your cycle is resetting, this might be a good time to look deeply, connect with yourself, and shed anything from the previous month that doesn't serve you anymore and spring forward a tad lighter.

Follicular Phase: Inner Spring

The follicular phase is like the spring when energy and possibility are in the air. As estrogen begins to rise (... and a whole lot of other things happen physiologically), you may notice that you're a bit more resilient. Your energy might not suffer as much if you don't sleep well, or you may not get hangry as easily if you skip lunch. This doesn't mean you *should* skimp on sleep or nutrition! It simply means that you might be able or inclined to do more on less compared to other parts of the cycle.

From a TCM standpoint, this is a time when you're moving from the Yin part of your cycle to the Yang part. Similar to the snow melting, sun shining, buds sprouting, and weather changing, this can be a time of transition and growth. You may feel more energetic, focused, and full of possibility and hope.

This is typically when many menstruators report feeling their best, so you can avail this energy to increase the tempo a bit. It may be a good time to implement a change you've been meaning to make or to start something new, like a health habit.

Ovulation: Inner Summer

The time around ovulation is like the summer or full moon. From a TCM perspective, this is the most Yang part of the year and cycle, when energy is abundant. Yang energy is more masculine in nature; it has more to do with taking action versus reflecting and resting.

Estrogen and progesterone both rise shortly after ovulation. You may notice that you have the capacity to take more on. As a result, you might do more, be more social, find it easier to say yes, share your ideas with the world more openly, and do the things that need to get done.

That said, it's still prudent to pace yourself! The point is not to use up all your stores in the spring and summer when energy is high(er). You want to get things done, but balance it out with play and recovery–perhaps play and recovery look a bit different in this phase.

Without balance, you might just find yourself depleted later in the luteal phase or during the period ... and if it's not immediately after, then it might be down the road. I think we all know that we cannot, with sound mind and body, go-go-go all the time and not have some sort of reckoning later. Remember: there is a Yin to the Yang of it all.

Luteal Phase: Inner Autumn

This part of the cycle is characterized by a rise in hormones followed by a drop, i.e., the late summer transitioning into autumn. In TCM, we're transitioning from the high Yang energy at the beginning of this phase into the Yin part of the cycle as we near a period.

As you already know, we all manifest these shifts differently. The early luteal phase seems to carry with it some of the energy from ovulation time until the late luteal phase when hormone levels begin their downward journey. You might notice that you need to slow down a bit–perhaps stack your easier or lower-effort tasks and reflect on the few weeks that just wrapped up. There may be some fatigue, metabolism and/or appetite change, and fluid retention.

There might also be a lot of feelings, including dread (very understandably in many cases due to looming severe premenstrual disorders or menstrual conditions). However, our feelings are often influenced by the ideas we have of how we think we *should* feel, which, in turn, are influenced by social norms (that tend to be of the same-way-every-day kind). Moreover, many of us don't know where we are in our cycles in the first place, which means we're typically reacting to hormone shifts without acknowledging their existence.

I wonder if we would experience a higher degree of ease or lesser degree of symptoms if we went into the luteal phase more mindfully, accepting that this change is neither good nor bad–it simply is. Maybe the downshift that often accompanies this part of the cycle is an invitation to do something slightly different.

We have some data to back up that when we share our experiences with our partners, when body literacy improves, when we talk about periods more neutrally, and when we slow down, the premenstrual time is smoother. This seems in line with that.

As such, honouring this part of your cycle could look like leaning into comfort, working on a more neutral or positive mindset, taking things off your plate, connecting with loved ones, and saying no to non-essential tasks. Consider carving out more time to take care of your wants and needs (by moving, doing your self-care, eating enough, being more mindful, etc.) before the menstrual cycle starts afresh.

What Does It All Mean?

Essentially, this is another lens to look at your body and life through–a lens that many of us are unfamiliar with.

Not everyone will follow the patterns laid out here (and that's okay!)–there's no one "correct" way to have a menstrual cycle. We all experience our menstrual cycles differently, and that experience will change throughout life.

You might not have a menstrual cycle regularly or at all depending on whether you're in perimenopause, have a hormone-based condition of some kind, no longer have ovaries, or are using hormonal contraception.

Body literacy is pretty much always relevant regardless of your menstruating status. If you're up for it, gather some data and see what *is* happening for you. Pay attention to the ebb and flow in your body, and see if you can and want to lean into that cyclicity to move toward feeling better. This might be an opportunity to explore a different way to schedule and prioritize your day-to-day.

The term "cycle syncing" (with a corresponding biohacking methodology) was coined and trademarked in 2014 by Alisa Vitti.[27] It describes a twenty-eight-day female infradian rhythm based on the menstrual cycle and uses different lifestyle practices–nutrition, exercise, supplementation, scheduling, mindset

---

[27] *Vitti, Woman Code, 2014.*

tools, etc.–for each of the four phases for better general menstrual health as well as hormone-based conditions.

Since then, the online wellness space has exploded with different iterations of syncing lifestyles to menstrual cycles. The prevalence of this kind of information has facilitated awareness about the menstrual cycle and normalized the ups and downs of our hormonal ebb and flow, helping many people regain agency over their hormonal health. It has also allowed many menstruators to rekindle their relationship with their bodies as well as cultural and traditional practices. I love this!

Some approaches are fairly broad, while others are expensive and complicated with a focus on periodizing all aspects of life including different supplements for each week of the cycle.

Remember, people are also trying to sell you things. You do not need a different supplement for each week of the cycle. Many syncing approaches demonize cortisol and estrogen, while they liberally extrapolate hormone research to back claims as well. Truth is, we don't have robust data (yet?) on changing up nutrition and exercise weekly based on the different phases of our cycles; time and further research will help us elucidate more of the nuance in these areas.

This doesn't mean that you can't periodize some of your habits–you do you! You may need extra support premenstrually if you suffer from a premenstrual disorder, or additional strategies if you have debilitating menstrual symptoms on top of your medical management strategies. If you have cyclical traditional or cultural practices pertaining to different phases of the cycle, implement those to connect to your roots. That's obviously different.

For general menstrual wellness, we actually don't need to do too many things differently if our foundations are in place, like sleep, rest, exercise, nutrition, and connection.

Some of the benefits of syncing your lifestyle to your cycle actually lie in the body awareness it creates: you start noticing what's happening from week to week. It normalizes some of the changes that you experience in different parts of the cycle, like fluid dynamics and metabolic variability. It puts the focus on adding nutritious foods like berries, green veggies, seeds, fibre, and plenty of water while encouraging you to decrease hyper-palatable foods low on the nutrition scale like cookies, chips, and soda. It reminds you to slow down, to implement self-care, to be more intentional in life, and to deepen your connection to your body and others. This is all great! Knowledge, education, normalization, solid nutrition, and regular self-care are the foundations of good hormonal health even if you don't change them up cyclically.

Having said that, I do want to add some nuance to a few syncing approaches I've seen online. The first has to do with nutrition. Not much needs to change week to week on the nutrition front if you are already eating enough and your nutrition is balanced–perhaps a bit more water and fuel in the luteal phase is where your focus needs to be, especially if you're active; resting metabolic rate can go up slightly in the second half of the cycle.

Decreasing coffee and alcohol might be another consideration. Other than that, general food and nutrition can stay similar throughout the cycle (check out Chapter 29 for more on that).

The second has to do with exercise. I used to buy into the switching of exercise types with each phase of the cycle, especially decreasing exercise intensity in the luteal phase. It seemed like a cool thing to do. I've changed my tune on this thanks to exercise scientists sharing research and stats on female exercise and delving into the research myself.

Most females–more than 54 percent in Canada–do not meet minimum exercise requirements which are set at 150 minutes per week.[28] This means that we need to encourage and get more females (and people in general) moving. There's no replacement for exercise; regular exercise–specifically moderate- to high-intensity cardio and resistance training–is the lifestyle factor that has irrefutable evidence for metabolism and heart health, many period problems, cognition, strong bones, mental health, and longevity.

[28] *Statistics Canada. 2019.*

Discouraging menstruators from engaging in formal cardio or resistance training while doing gentler exercises for seven to fourteen days of the month is a disservice to us. The more we can build our muscle and condition our cardiovascular system while estrogen is around in our cycling years, the better it is for our long-term health.

Fuel, rest, self-care, sleep, and hydration are typically the factors that impact your exercise, mood, performance, and hormonal symptoms more than anything else in the luteal phase. Yes, hormones are changing, but it doesn't necessarily mean that we need to do a whole lot that's different; the basics still apply. You could consider breaking up your workout or decreasing the duration of your workout–do whatever is best for you!

If premenstrual symptoms or period pain are getting in the way, then seeking care can help you with your quality of life. But, don't skip the formal exercise. I know it's easier said than done, but sometimes we have to do things even when they're hard (reminder: you can do hard things!). Chapter 30 has specific details on exercise including cardio versus resistance training.

Syncing your lifestyle too meticulously is not particularly practical and not always relevant. I encourage you to take more of a bird's eye view and approach your menstrual cycle from a place of curiosity. As discussed earlier, not everything needs to be tampered with.

Cyclical living, to me, is broad in its scope. It embodies the concept of pacing yourself. It's understanding that there are shifts over the menstrual cycle, and you can't push through everything without redress at some point down the line. It's about cultivating an awareness of *your* body and what it's doing over the monthly cycle (and longer term as well).

Then, you can add in support when needed, not blow through all your reserves when you're feeling well, take it easy when you're feeling a bit low, replenish on an ongoing basis so that all parts of your cycle are supported, celebrate what's going well, and/or not change anything at all.

It is not about constantly changing things up, walking on eggshells, or not pushing yourself out of your comfort zone; it *is* about balancing out times of effort with rest and recovery.

It's not dissimilar to your personal financial cycle for a certain interval. You have to assess what your baseline is, track your income and expenses, find ways to improve, see where you're doing well (and celebrate that!), account for times when expenses will be higher, save or coast when you're doing okay, plan for emergencies, gauge your relationship with money itself, and consider getting some help for clarity and strategy. And, it's about setting yourself up well for the future.

CHAPTER HIGHLIGHTS

- Cyclical living is all about noticing the intricacies of your cycle and leaning into what's happening to go with the flow.
- The parts of the menstrual cycle have been likened to the seasons of the year. They all are dynamic and interdependent parts of a whole that do not exist in isolation.
- Your period is the winter of the menstrual cycle when rest and building up your stores are good ways to spend your time. It is also the most Yin part of the cycle according to TCM, which is when nourishment (physical as well as mental/emotional) should be prioritized.
- The follicular phase is like springtime when energy and possibility are in the air to try new things and get things started.
- Ovulation is like the summer, and it's the most Yang part of the cycle. Yang energy is about taking action and getting things done.
- The luteal phase is like autumn, with the transition from the high energy of the summer to the slower days of winter. This is a great time to start focusing more energy inward and spending time ending the seasonal (and menstrual) cycle on a reflective note as a new cycle approaches.
- Cyclical living is about pacing yourself with the understanding that one phase of your menstrual cycle will lead to the next, and one whole cycle will impact the next.

CHAPTER 29

Nourishing your Cycle

Most of us learn about food and nutrition from our family of origin, school, books, media, and online sources. We're basically left to our own devices when it comes to synthesizing all of that for daily consumption (I like my puns!).

Unfortunately, the conventional medical education system does not emphasize nutrition enough, which can lead to doctors being unequipped to guide patients and sometimes even dismiss the role of nutrition in health.

Then, there are wellness gurus who tout that diet (which can even include restrictive regimens and the promotion of disordered eating) can heal everything under the sun. The aphorism "you are what you eat" has been taken disproportionately out of context.

The reality lies somewhere in between: nutrition is an essential pillar of health and well-being, it's an important modifiable factor for chronic disease, *and* it stands alongside other foundations.

We have many great food and nutrition studies that have been replicated and corroborated over time. The good news is that there are themes associated with promoting health, lowering all-cause mortality, and decreasing the risk of non-communicable illnesses. Moreover, there isn't one "perfect" diet; there's room for flexibility (more on this later in the chapter).

The bad news is that tried-and-tested strategies don't usually get sensationalized headlines, so it's easy to get caught up in the minutia that viral headlines and clickbait offer, especially in an age where everyone has an opinion and is sharing it loudly.

It's easy to get tangled in the web of dogma, conflicting information, and varying opinions that all seem to be backed by some sort of cherry-picked research. I know that I have fallen into the trap of dietary fads and erroneous information personally and professionally.

So, when you're browsing Google results and social media, try to anchor yourself in healthy skepticism. Paid advertising affects what you see. Cherry-picked and/or poorly conducted research can make it seem like a dietary pattern is evidence-based; however, we need to keep the big picture in mind and not focus on the outliers.

Diversity, debate (not trolling), and different perspectives are helpful to learn, unlearn, foster curiosity, and continue to grow our understanding of health and nutrition.

BOX 21. Cognitive biases

There are numerous kinds of biases that affect research, interpreting data, and walking through life.

Confirmation bias has to do with favouring and interpreting data in a way that supports your existing beliefs, understanding, and values, typically ignoring information that is contrary to that.

We're all pretty much prone to this, especially when it comes to emotionally charged topics like health, politics, relationships, and ethics.

This may play out when you're looking for or interpreting information. It can even affect the kind of information you recall from your memory. Your memory isn't perfect to begin with; it's kind of human to be selective about the memories we use to back up a topic.

Confirmation bias can play out at the individual level, but it has systemic effects as well.

Someone who thinks that all sugar is bad might negate any and all of the evidence and recommendations we have for eating fruits as part of a healthy diet to lower the risk of chronic disease is an example.

A doctor who has diagnosed you with IBS may not ask any questions about your menstrual symptoms of pain and midcycle spotting, completely missing your diagnosis of endometriosis, is another example.

It can be difficult to let go of our beliefs and take in data in a more detached manner. Emotions and defences can go up faster than lightning. Relationships can be torn apart, even.

Coming from a place of curiosity, being open to changing your mind when presented with new information, and broadening your understanding using different perspectives helps with growth and discernment.

We need more of this in all areas of life. This is how to take better care of ourselves, our planet, improve standards of living, create better infrastructure, and move forward as a whole.

Diet Culture, Beauty Standards, & Weight Stigma

The messaging from diet culture, patriarchal beauty ideals, and capitalism revolves around shrinking ourselves and having us spend our time pursuing an image of beauty that is always changing and not inclusive. Oftentimes, our families, healthcare providers, filters, body-tuning apps, celebrities, influencers, and the media all perpetuate this messaging. It is no wonder that many of us have torturous relationships with food and our bodies. I don't think I know anyone (myself included) who is immune to the effects of diet culture and toxic beauty standards!

Systemic inequities and access to resources play significant roles in how health plays out as discussed in Chapters 22 and 23. Yet, "somehow," we have reduced the complexities of health to one number; a number that so many people attach their whole self-worth to: weight.

Sizeism and fatphobia are pervasive in our communities: from a lack of consideration in infrastructure to how fat people are treated in society. The healthcare system has particularly been susceptible to this with the "war on obesity". Weight stigma is rife in the medical system and poses an enormous barrier to millions of people accessing care.

BMI (body mass index), a tool that was designed as a statistical measurement for groups–not individuals–has taken over healthcare like nobody's business. Instead of having proper conversations about a patient's health habits, the BMI is used to direct treatment plans even when patients are coming in for something entirely unrelated.

Doctors sometimes have patients waste their precious time and efforts on weight loss to fit the "normal" BMI category even though these patients have within-range cholesterol, glucose, and blood pressure readings, feel well in their bodies, and have lifestyle components like exercise, stress management, purpose, wholesome nutrition, community, etc. in place. Let's not even get started on the recklessness of many recommendations to lose weight without screening for disordered eating patterns or an outright eating disorder, metabolic and gut dysregulation, substance abuse, and even suicide.

Yes, there is a link between carrying excess weight around our organs, increased chronic disease risk, and poor health outcomes. Yes, we need to encourage more people to engage in physical activity and eat more nutritious foods. However, there are other ways of measuring body composition than BMI, there are many metrics of health we can focus on instead of someone's gravitational pull to the earth alone, and there are ways to support patients looking to improve their health beyond recommending outright weight loss especially when we know how unsuccessful and damaging attempts can be.

Weight is a product of a multitude of factors (no surprise there, I hope!) like genetics, family of origin and upbringing, medication use and health history, hormone signalling, gut alterations, metabolic impairment, income, and access to care and resources.

There is so much more to all of these topics that is beyond the scope of this book. I encourage you to do some reading on these subjects.

The association between thinness and virtue and that between fatness and laziness is a narrative that many people hold (just look at a comments section under a fat loss post online, and this will become extremely apparent ... or, you can look at all the research studies on the topic if you want the science). This spills over into food and nutrition too, whereby many of us assign morality to foods and behaviours: good vs. bad, clean eating vs. cheating, even though food isn't inherently good or bad. Amongst other things, this rhetoric contributes to a dysfunctional relationship not only with food as a whole but also individually with different types of foods.

In my patient population, I see a lot of disordered eating and body dissatisfaction: from yo-yoing and crash diets to bone loss and suicide attempts.

I personally was exposed to diet culture and toxic beauty standards at a young age—as are most people—with relatives tearing apart their own and other people's bodies constantly (hello, South Asian "norms" where a conversation isn't complete without a comment on your weight, how fair or dark your skin is, and how you look overall ... I know, this isn't only found in South Asian societies.).

The more I learn about diet culture, the more I try to understand where patients are coming from so I can support them better myself and through resources and appropriate referrals to mental health professionals, especially culturally appropriate ones. I'm no psychologist, but working on this–something that is vital to surviving and thriving–is crucial. Body size and food issues can take up a lot of mental and emotional space, often being a life-long battle.

We need food to survive; we need adequate food for healthy menstrual cycles. For those of you needing the reminder: weight is not the ultimate metric of health or your worth. You are not your weight.

Nutrition is mostly about health and living, not weight. So, that's what we'll focus on for the rest of this chapter.

The Balancing Act of Food and Nutrition

One of the most frequent recommendations I give to patients around food is to eat regularly. This seems like a fairly simple point. However, food restriction is so commonplace that it can be often difficult to recognize (note: Part Five of this book contains commentary on food accessibility).

Let me paint a picture of what I often hear in my practice:

You're a person with a lot going on in life. You typically skip breakfast or sometimes have a cup of coffee and a banana. Then, because your schedule is packed, you have soup or a small salad for lunch. Before you know it, it's after 4 pm. You're tired and hungry, so you reach for whatever you can get your hands on. You then have your first (and only) proper meal at night time. After dinner, you get a craving or you feel peckish, so you have a bit of a snack attack. After you nosh, you wonder why you're hungry ... because, *technically,* you ate three times. You berate yourself for not having the "discipline" to stop eating at night. You go to sleep vowing that you'll do better tomorrow. And then, it starts all over again.

First, you're not alone.

Second, this is a classic case of not eating enough when your body is most active, i.e., during the day after a night of fasting, when you're doing all the things whether you're at work or school or doing everything else that life requires of you.

The fact is that your body needs fuel. If you don't eat enough during the day, you'll likely receive hunger signals later in the night. What will also probably happen is a mid-afternoon crash, irritability (hangry, anyone?), headaches, blood sugar dysregulation, night-time cravings, an impact on your metabolism and hormones, feeling like you lack control over your appetite, feelings of failure, and much more.

If this list reminds you of PMS symptoms, then you'd be correct in thinking that eating regularly helps with some of the symptoms experienced before your period. From Chapter 16, we already know that eating inadequately in the luteal phase can worsen PMS, partially due to increased metabolic needs.

You don't need a dietary overhaul. You need someone to have real conversations about bodies and nutrition with. Food is the gas in the tank for stable energy and healthy hormones, with the not-so-surprising effects of better appetite and mood control. My patients are often surprised at how much of a difference in energy they feel when they *actually* eat breakfast ... and their other meals. They begin to better understand the effect food, or the lack thereof, really has.

Now, you might be wondering what exactly to eat for breakfast. Well, here's where the other aspect of balance comes in: the composition of your meals. You need all of the macronutrients–protein, fats, carbohydrates–to have healthy menstrual cycles.

A balanced diet is one through which your nutritional needs are met. This not only gives your body the resources it needs to function well but for satiety and blood sugar regulation as well.

In the last decade, there's been a big shift of moving away from carbohydrates with the ketogenic diet becoming popular for weight loss. The thing is: carbs are an important source of fuel. Moreover, foods rich in carbs are also the foods we get fibre and antioxidants from (more on that later). When choosing carbs, focussing on grains like wheat, rice, oats, and millet as well as lentils, beans, veggies, and fruits is the way to go for your health (versus hyper-palatable unprocessed foods).

BOX 22. How much protein do I need?

Protein requirements differ depending on who you are, how much activity you engage in, and your overall health.

The Canadian guidelines currently recommend 0.8 grams of protein per kilogram of body weight. So, if you weigh approximately sixty-five kilograms (i.e., 143 pounds), then that comes out to fifty-two grams of protein a day. This might be a good place to start for most people, especially if you're sedentary or only mildly active.

Studies show that aiming for 1.3 to 1.6 grams of protein or over per kilogram of weight is better to maintain and grow muscle, protect against age related bone and muscle loss, support recovery, and help with satiety and blood sugar.

When it comes to fats, the dietary pattern associated with better heart, hormonal, cognitive, and metabolic health while decreasing the risk of cancers and other non-communicable chronic diseases includes adequate omega-3s and mostly plant-derived fats, especially olive oil, while limiting saturated fat intake.

Let's get into that a bit further. Omega-6 fatty acids are usually easy to get through our diets through all kinds of plant foods, especially cooking oil. Seed oils often get a bad rap online as being inflammatory; however, these claims are mostly unsubstantiated and primarily anecdotal. In fact, there is a decrease in the risk of heart disease when we replace animal-based saturated fat with omega-6 oils.

Omega-3s, which are commonly associated with fish but are produced by algae, are important for overall health. Walnuts as well as flax, hemp, and chia seeds are plant sources high in omega-3s. You may need a higher quantity of omega-3s for a specific health concern like period pain, eye conditions, and cardiovascular disease, but regular consumption of foods high in omega-3 is sufficient for general health, including your joints, brain, and periods.

The kinds of fats we want to limit in our diets are saturated fats–these are solid at room temperature. Dietary patterns with high consumption of saturated fat like red meat and full-fat dairy (paired with low consumption of mono- and polyunsaturated fats typically found in plants) have been associated with an increased risk of colorectal cancer, heart disease, cholesterol problems, PCOS, fatty liver, and endometriosis.

What can you do to decrease saturated fat? Well, consuming red meat less than two to three times per week is a good place to start as we see higher morbidity rates above this intake level. Add more plant-based protein and fat sources, consider switching to low-fat dairy, and focus more on lean cuts of animal protein if and when you eat meat.

You need dietary fat and you need to fuel your body in general. If you go too low on calories, your menstrual cycle can get disrupted. The threshold for this varies from person to person. When resources are low, reproduction is not a priority. Your body will divert resources toward survival.

You need energy, aka food, to function and feel well. Instead of removing entire food groups, focus on balance and adequacy. Incorporate lean proteins like beans, carbohydrates like quinoa and whole wheat, healthy fats through nuts and fish, and add lots of veggies, herbs, and spices, when you're planning and cooking meals most of the time.

BOX 23. Examples of nutritionally-balanced meals

- Classic combo in numerous parts of the world: rice and beans or rice and lentils with a side of veggies (daal chawal is my go-to comfort food)
- Stir-fried veggies tossed with a protein of choice, peanut sauce, and noodles
- Salad: greens, other veggies of choice, protein of choice (mine's edamame), quinoa or potatoes, dressing made with olive oil, vinegar, and herbs
- Wrap: protein of choice, greens, other veggies of choice, and a dressing with nuts or seeds like hummus or tahini
- Smoothie: berries or seasonal fruit, protein powder, nuts or seeds, greens, pinch of cinnamon, liquid of choice

Another aspect of balance might include the concept of hot and cold foods. In South Asian cuisine, the balance between the hot and cold nature of foods is important for digestion. Other parts of the world have similar views. In Traditional Chinese Medicine, warm and cooked foods are supportive of digestion.

There is no shortage of recipes online that you can use or draw inspiration from. Include tastes and foods that you enjoy, that are part of your heritage, and that meet your other preferences.

Consume plentiful plant foods

Plants offer a wide array of nutrients important for health. In its latest update, the Canada Food Guide finally upped its game to incorporate the latest evidence-informed recommendations that focus on plentiful fruits and vegetables, whole grains, nuts and seeds, and plant-based protein sources.

A major nutrient that we get from plant foods is fibre. Fibre only comes from plants. Unfortunately, most North Americans do not meet the daily recommended intake of fibre. The recommendation for adults is twenty-five to thirty grams per day.

Not only does fibre facilitate waste elimination via bowel movements, it also helps with satiety, a more diverse gut microbiome, healthy cholesterol levels, and proper blood sugar management. Low fibre intake has been associated with increased digestive issues and PMS symptoms in the luteal phase.

BOX 24. Twenty-five grams of fibre a day

There are many ways to combine plant foods to hit your fibre target for the day. You could even add in psyllium husk (which is what Metamucil® is).

- Half cup cooked beans (eight grams) + two tablespoons pumpkin seeds (five grams) + one apple (five grams) + one cup cooked brown rice (six grams)
- Two pieces of multigrain toast (four grams) + handful of nuts (five grams) + half cup berries (four grams) + salad made of one cup raw spinach, quarter cucumber, half tomato (six grams) + one cup whole-wheat pasta (six grams)
- Half cup uncooked rolled oats (four grams) + one pear (three grams) + 1 cup cooked lentils (fifteen grams) + half cup cooked broccoli (three grams)

If you were to eat two pieces of whole-grain toast, an apple, a big salad with some nuts or seeds, and some rice with lentils or beans in your day, you'd hit your fibre target. It sounds simple, but a majority of North Americans consume far below the suggested intake.

If your fibre intake has historically been low, adding in fibre-rich foods slowly over time is a prudent way forward as your digestive tract and gut microbiome adapt to that. Certain types of fibre can be quite fermentable, meaning that you could experience bloating or change in stools in the process of adding more, especially if you have a digestive disorder like IBS.

The recommended intake for fruits and veggies is around six to ten servings daily. One serving can be:

- half a cup of berries
- one medium-sized carrot or apple
- one cup of raw leafy greens

Eating the rainbow is a good rule of thumb to get a diversity of nutrients and antioxidants. Low fruit and vegetable intake has been associated with more PMS symptoms as well as painful periods and endometriosis. Although antioxidant-rich berries, pomegranates, cherries, and cruciferous vegetables (like broccoli and cabbage) have some added benefits to hormonal and overall health, having *any* fruits and veggies that you like and can tolerate is a good place to start. One serving per day is better than zero!

Smoothies can be a great way to pack a lot of plant power into your day. I'm not a morning person, and a smoothie is what I have for breakfast on most days. My husband and I can make our smoothies and clean up after ourselves in under five minutes. My go-to recipe includes frozen blueberries and cherries, greens, hemp and flax seeds, vegan protein powder, and a mix of soy milk and hot water.

Frozen fruits and vegetables are a convenient way to get your daily servings in. I keep frozen berries, mixed veggies, as well as legumes like edamame and peas in the freezer because it makes eating vegetables easier. Frozen veggies tend to also be budget-friendly (especially if you buy in bulk or when on sale) and quite versatile–throw them into soup, pasta, porridge, smoothies, or rice. They are perfect for when you've made your way through your fresh produce, as you inch closer to your next grocery shop, or maybe when your fresh greens have wilted because your aspirational salad didn't come to fruition (I speak from experience!).

Plant foods like soy, lentils, and beans are a great way to add in lean protein. The updated Canadian and American food guides have both encouraged eating more plant sources of protein, especially in place of fatty cuts of meats.

BOX 25. Is soy safe?

Soy, as long as you don't have an allergy, is not a food to avoid even though its tarnished reputation has followed it around for years.

Soy contains phytoestrogens that help with PMS and hormonal changes in the menopause transition. Many studies have linked soy consumption to reduced risk of cardiovascular disease, vasomotor symptoms of perimenopause, and many cancers. It also has been shown to have no effect—positive or negative—on blood levels of hormones.

So, go ahead and enjoy your tofu, tempeh, or soy milk. Soy milk tends to be higher in protein compared to other plant-based milks if you're someone who avoids animal dairy. In fact, soy is a great source of lean protein.

Shelled edamame beans are a freezer staple for me. Offering around fourteen grams of protein per half cup, I add them to salads and noodles often. They also make for a great snack with a bit of salt, pepper, and red chilli powder.

Another way to incorporate some plant power into your daily regimen is through culinary spices and herbs. For specific conditions, herbs and spices in culinary quantities might not offer therapeutic benefits; however, they can provide general support as well as antioxidants, micronutrients, and comfort. Some of my favourite herbs and spices, especially with my Pakistani background, include turmeric, garlic, ginger, cilantro, chilli pepper, cardamom, and fennel.

Eating more plants, whether you are plant-based or not, is always a good idea. As someone who has been living the vegan life for over seventeen years, I'm a big advocate of adding in more plant foods, especially plant-based proteins, not only for health but also for the environment and the animals.

Stay Hydrated

As the Canada Food Guide says, "Water should be the drink of choice", and I could not agree more.[29] It doesn't have to be eight glasses per day. This number is a bit arbitrary and not rooted in science. It's not a bad target, though. Water requirements vary depending on your size, activity level, age, salt intake, fruit and veggie intake, what's happening in your health and life, weather, and more.

There is a correlation between dehydration, period pain, headaches, and lethargy, so those might be reasons to increase water intake.

Many people struggle with drinking water regularly. We get a bit of water through the food we eat. A good strategy could be to fill up a big water bottle or jug and aim to finish it by a certain time of day. There are apps that you can use as well. If you don't particularly care for water, herbal teas, fruit infusions, and soups are also ways to get hydration in.

As you're adding more hydration, avoid or minimize the intake of sugary drinks and alcohol. In case you missed it, the Canadian Centre for Substance Abuse and Addiction's guidelines for alcohol consumption have (finally!) changed in 2023 to reflect that not drinking is the best practice. Anything over three drinks per week starts to increase the risk of developing cancer (for e.g., breast and colon), heart disease and stroke, and liver problems. So, sticking with water is the best option.

Slicing Through Stress in the Kitchen

Another facet of eating is the overall balancing act of buying groceries, staying within your budget (hello, inflation and profiteering!), deciding what to eat, preparing for the week ahead, and the effort all of this takes.

I don't know about you, but I eat a lot of the same food regularly: daal and rice, tofu and veggie stir-fry, tacos, burrito bowl, ramen, protein smoothies, pasta or

---

[29] *Government of Canada. Canada's Food Guide. 2021.*

roti with protein and veggies. Sometimes, I'll cook something new, but those are my go-tos.

The fresh ingredients for these foods are always on the grocery list, and we replenish the dried and bulk ingredients on an as-needed basis. I also have my go-to snacks, which usually are some kind of fruit, dark chocolate, popcorn (with nutritional yeast!), carrots, and protein bars. I always keep a few things in the pantry and freezer, like canned chickpeas and frozen burgers respectively, for the inevitable days when I don't want to put much effort in. I also eat out!

We eat every day, multiple times per day. If you often wonder: "What should I make?", "What should I eat?", "I'm too tired to cook," and "I wish what I want would magically appear in front of me", you are not alone.

Thinking about food takes up a lot of time and space, especially if you're accounting for other people in your household. Similar to me, you probably have the routine meals that you make time and time again, the places near you for ordering in, and snacks that you reach for regularly.

This offers a good place for you to start making your life simpler. Use this data to create a grocery list template, or to pre-load your online cart. Then, you can add or subtract things as you go.

Consider preparing a few things ahead of time for when you're busy. Meal prepping can be helpful. You can make a giant batch of chilli or soup for a few meals during the week. You can make overnight oats for the next few days.

You can add a dash of flexibility to your meal prepping, and have it not be the kind where you have full meals prepared and sectioned. Maybe you bake your protein for the week so you can use it in a sandwich, wrap, or salad for lunch. Maybe you cut up veggies just for the next couple of days. It's supposed to make life easier, so figure out what works best for you and involve your household if applicable.

Like everything else in life, you get better at this sort of thing by rinsing and repeating. The same thing goes for cooking–you get better at cooking by actually cooking. Start with one recipe and move on from there. Making your food is one way to have agency over your nutrition while ensuring that you're not consuming too much sodium or saturated fat.

On the flip side, there's also a case to be made for convenience and ease. Ordering in has never been less complicated. With the click of a button, you can have anything delivered straight to your door. Try to use the same principles of balanced nutrition when eating out or ordering in (most of the time, anyway).

BOX 26. What about ultra-processed foods?

Ultra-processed foods (UPFs) are categorized based on the NOVA Food Classification System as:

> "made entirely or mostly from substances extracted from foods (oils, fats, sugar, starch, proteins), derived from food constituents (hydrogenated fats and modified starch), or synthesized in laboratories from food substrates or other organic sources (flavor enhancers, colors, and several food additives used to make the product hyper-palatable)." [30.]

Not all UPFs are made equal. There is a wide range of foods that fall in this category from cola and pretzels to fruit yoghurts and tofu. As you can tell from those examples, the former two aren't in the same tier as the second two.

For health promotion, we're usually talking about reducing the hyper-palatable kinds of UPFs (commercially available cookies, candies, pastries, chips, sweetened juices, etc.) weighing low in nutritional value. Frequent consumption of these foods is associated with many chronic diseases especially since they also tend to replace other foods that offer more in the way of nutrition.

Do I want people to consume these regularly? No. Might it be best to avoid them completely? Sure. Would it be better if we made or ate snacks offering

---

[30] *EduChange and NUPENS, 2018*

better nutritional value? Yes. But, is it okay to have flexibility around these foods, enjoying them from time to time? Yes.

These are the same foods that typically bring up guilt and shame because of their links to being "bad." They are also hyper-palatable with a significant amount of sugar, fat, salt, and carbs; this makes you want them more.

We don't want these foods to form the bulk of our diets. However, a sprinkling of flexibility to create room for enjoyment, tradition, nostalgia, and convenience needs some consideration. Our bodies are resilient. I like to periodically enjoy salt and vinegar chips, Slims Potato Sticks (if you know, you know!), and Sour Cherry Blasters.

Food restriction, especially in the name of health, has become extremely common. Every few years, there's a different food you "must" have or "need" to avoid. Confirmation bias plays a major role in our nutrition these days!

One of the biggest trends of the 2010s was gluten elimination; avoiding gluten is the action to take if you have Celiac disease as gluten wreaks havoc in your gut and body due to an autoimmune response. Not everyone needs to avoid gluten willy-nilly, though. Some people have a gluten sensitivity. Others may be reacting to certain (but not all) gluten-containing grains. Some might be reacting to ingredients that they usually have with gluten. Some might be reacting to the fibre or fermentable sugars that gluten-containing grains have. And, some people make multiple dietary and lifestyle changes at the same time which makes it difficult to discern what is helping and what is not. There are many possibilities here. You can experiment if you'd like, but I don't think it's good practice to simply eliminate all gluten just because it has been labelled as the root of all evil by some.

Your food and nutrition habits don't exist in a bubble–they need to be taken in context. There is a role for therapeutic diets when clinically indicated for certain concerns, but restriction is not the way–flexibility is. Nutrition seems to have a cumulative effect over time. A few meals or snacks aren't going to make or break your health.

Even if you're working with a practitioner like a nutritionist or dietitian, the plan needs to be rooted in practicality; we can't simply toss out all the foods that bring you joy and pleasure, that might not be the most nutritious but help you stay on track with other aspects of your health, or are part of your heritage and culture.

Alicia Elliot put it beautifully in her book, *A Mind Spread Out on the Ground*, "Food that carries the love of our ancestors *can* be medicine… an act of absolute, undiminished intergenerational love."[31] So, please, don't remove foods from your heritage because diet culture labelled them "unhealthy" or because you read anecdotal evidence online.

White rice is a good example of this. It is a big part of the culture and myriad traditional dishes across the world, typically paired with a protein like beans, lentils, or meat. It can be unfathomable to even think about restricting it.

Brown rice is often recommended as a substitute for white rice. Brown rice has a tiny bit more fibre, vitamins, and minerals. Since rice is typically paired with a protein like beans, lentils, or meat, we don't necessarily need to worry about the marginal differences between white and brown rice, especially when there are other (more impactful) changes we could be making.

I love rice, especially my mum's biryani. I will never have biryani with anything but white basmati rice unless my hand was forced! Food is a source of connection, pleasure, nourishment, and community. Take what you consume online with a grain of salt!

Get Rid of Your Crap

"TMI (too much information)" … that's what many patients say before disclosing their bowel habits.

The thing is: that it's not TMI. I think we need to talk about this kind of crap more!

---

[31] *Elliot, Alicia. A Mind Spread Out on the Ground (p. 116). 2019.*

Puns aside, your gut is important not only to your hormones, but also to your mood, cholesterol, and overall health.

There are many steps in the digestive process. Digestion starts before you even put food into your mouth. Smelling, seeing, and salivating start the digestive process. When you eat in a hurry or while scrolling through email, your parasympathetic nervous system isn't engaged.

The parasympathetic nervous system is the "rest and digest" or "feed and breed" part of the nervous system, whereas the sympathetic nervous system governs the "fight, flight, or freeze" mode. Eating in a relaxed environment so that your parasympathetic nervous system is engaged is always a good call. It allows you to engage your digestive system fully. As a bonus, it's good for your mental health.

Your gut microbiome is another crucial piece of the digestive process. This doesn't necessarily mean you need a probiotic. Your gut microbiome is constantly changing in response to your day-to-day nutrition and life. The bacteria in your gut play roles in your immune system, in the production of your feel-good neurotransmitter called serotonin, and in the absorption of nutrients like vitamin B12.

Normal bowel function can vary from person to person, and things might change transiently depending on where you are in your cycle and based on what else is happening in life. The most common pattern is that of slightly slower-moving bowels before a period and softer stools with a period along with more fluid retention.

A diverse diet containing adequate fibre, stress management, medications, hydration, and immune status all play roles in your gut function. Here's what a well-functioning gut may look like:

- Daily bowel movements
- Feces are formed, soft, easy to pass, and painless. Watch out for undigested food (unless it's corn), chronic loose stools or diarrhea, bitty or hard stools, straining, bleeding, mucus, and pain

- Some gas production is a normal part of digestion, but it should not be painful or excessive
- Some fluid retention as you eat and drink throughout the day is expected

For our bodies to function well, for smooth periods, less PMS, and healthy hormones, we need to eliminate waste regularly, break down and absorb nutrients from our foods properly, and have a flourishing and diverse gut microbiome.

Constipation doesn't always look like poops that are MIA. You can have a small and bitty bowel movement daily and still be constipated.

Adequate hydration, fibre intake, herbs like marshmallow root and slippery elm, laxatives like magnesium citrate, exercise, deep belly breathing to activate your vagus nerve, stress management, and using a product or step to elevate your legs to have a more optimal rectal-anal angle (like a stool to rest your feet on can be helpful for bowel health).

There is a time and place for supplements and probiotics to support your digestion. Please don't waste your money on products you don't need.

Also, try not to spend all your time and effort shrinking the amounts of food you can eat needlessly. Undereating and food restriction have very real and vast impacts on mental and physical health.

If you suspect you are sensitive to something, i.e., it causes a headache, bloating, or acne, do a bit of trial and error before you cut it out of your life entirely. Better yet, talk to a healthcare professional. If you have a diagnosis like inflammatory bowel disease (IBD), irritable bowel syndrome (IBS), or gastroesophageal reflux disease (GERD), then discuss a treatment strategy with your healthcare team. The same goes for digestive issues of unknown origin; advocate for a comprehensive assessment and care.

Specific Nutrients & Strategic Supplementation

Supplements are supposed to do just that: be an accessory for your health. We can also use supplements as therapeutic aids in a more pharmacological manner, i.e., like a medication.

Supplements can include not only vitamins, minerals, and antioxidants, but herbal products as well. Nutrition and supplement considerations for general menstrual health as well as for different hormonal concerns are addressed in Part Four of this book.

Now, I'll briefly touch on a few nutrients and offer considerations as they pertain to menstruators, dietary patterns I see in my practice, and energy levels that impact the ability to stay on track with your health habits. None of this is medical advice.

- **Iron**: This was covered in Chapter 14. Iron deficiency is one of the most common nutrient deficiencies worldwide. Iron should be tested via a blood draw and addressed accordingly. Iron deficiency can lead to fatigue, poor focus, restless legs, anxiety, hair loss, shortness of breath, and heavy bleeding. Iron deficiency is common in menstruators regardless of the dietary regimen. If your iron is low, supplementation with approximately one hundred milligrams for at least eight to twelve weeks is recommended before retesting. As a menstruator, you lose blood regularly. Diet alone will most likely not bring your iron levels up if you're deficient. Any form of iron helps; however, if you have digestive issues with certain forms of iron, then the bisglycinate, polysaccharide, liposomal, or carbonyl forms of iron seem to be gentler on the stomach.
- **Vitamin B12:** This is another one that is testable through blood work. Like iron, B12 deficiency is also common, leading to fatigue, poor concentration, restless legs, tingling, and mood dysregulation. Some dietary patterns like plant-based diets may be low in B12, although many foods are fortified. As we age, our decreased ability to absorb B12 (and the lower food intake associated with ageing) can contribute to B12 deficiency.
- **Vitamin D:** The sunshine vitamin has been implicated in a variety of conditions like PMS, dysmenorrhea, and PCOS as well as generalized symptoms like pain and fatigue. Testing vitamin D is a matter of

running blood work, and gauging how much you need to take. Supplementation has never been easier with liquids, tablets, and gel caps being readily available at supermarkets and pharmacies. There is heterogeneity in vitamin D conversion in the skin, absorption of supplements, and conversion to the active form in the body, so those are things to keep in mind here. It is possible to have vitamin D toxicity if you take too much vitamin D, so please be wary. In Canada, the daily recommended dose for general health is 2500 international units (IU) per day.

- **Calcium:** I'm not sure what changed the emphasis on calcium for healthy bones in the last decade, but we need to make sure everyone's getting enough calcium regularly. Calcium is present in dairy products, non-dairy milk alternatives, nuts and seeds, veggies, and grains. Those assigned female at birth put on most of our bone mass by the age of twenty. By our early thirties, we've put on the maximum bone mass that we'll ever have. Then, it slowly starts going down, accelerating after menopause. This means that we have to protect our bones through adequate calcium, vitamin D, other minerals, and exercise. The daily adult recommendation for calcium is approximately 1000 milligrams, so make sure that you're getting approximately that regularly. As discussed in Chapter 16, calcium supplementation in the luteal phase can be very helpful for the physical and mental-emotional symptoms of PMS. Calcium is crucial for healthy pregnancy outcomes as well.
- **Magnesium**: This mineral, which is often considered a panacea in the wellness world, is becoming more and more deficient in our soils due to agricultural practices. It's also needed for hundreds of biochemical reactions, for helping with proper muscle function (think of headaches and cramping), for blood pressure regulation, for the digestive system, for the nervous system, and for sleep regulation. Testing magnesium isn't particularly useful in the blood. Supplementation may be helpful for the luteal phase (for headaches, cramping, and mood) as well as during the period, especially when combined with calcium. However,

too much magnesium (especially magnesium citrate or oxide) can exert a significant laxative effect.

- **Folic acid**: It is recommended by health organizations across the world that you get approximately 400 micrograms of folic acid if you are of childbearing age. This is important if you're planning to get pregnant, but also for accidental pregnancies, since folic acid is really important for cell division. A lack of this vitamin in early pregnancy can lead to birth defects amongst other negative outcomes.

- **Multivitamins and Prenatal Vitamins:** Multivitamins can be an effective way to get your minimum and recommended daily doses of vitamins and minerals. You can find multivitamins with vastly different ingredient lists, so finding one that is appropriate for your needs is important. It can be a great adjunct for people who are picky eaters, eat irregularly, have digestive problems, have undergone bariatric surgery, avoid certain foods, restrict food, or have low appetites for whatever reason. Prenatal vitamins are multivitamins specifically designed with a pregnant person and developing baby in mind. They typically contain lower levels of vitamin A and at least 400 micrograms of folic acid to support a healthy pregnancy. I support a fair amount of plant-based pregnant patients in my practice, and choline is an important nutrient I look for in a prenatal vitamin for this population (and anyone else who may not get adequate choline in their diet).

- **Omega-3s:** EPA (eicosapentaenoic acid) and DHA (docosahexaenoic acid) are the two omega-3s that are most useful for health along with their precursor molecule called ALA (alpha-linolenic acid). For general health, dietary omega-3 is sufficient through chia and flax seeds, walnuts, algae, and fish. For period pain, cardiovascular disease, dry eyes, inflammation, and cognitive health as you age, and if your dietary intake is low, then adding in an omega-3 supplement might be useful. The dosage will depend on your health needs, with at least 300 to 400 milligrams of DHA needed in pregnancy for the fetal nervous system.

I recommend talking to a knowledgeable practitioner like a naturopathic doctor before you take supplements to ensure that they're helpful and appropriate for you, to check for interactions, and also to weigh out the risks and benefits.

I take patients off supplements all the time! Nutraceuticals are a multi-billion-dollar industry, and I don't want you to be wasting money on things that don't *actually* serve a purpose in your health. Supplementation should have a purpose.

CHAPTER HIGHLIGHTS

- Nutrition is an essential pillar of health and well-being, but it stands alongside other pillars. You are not just what you eat–you are much more than that.
- When resources are low, reproduction is not a priority. You need to eat to function and feel well, and everyone has a different threshold for when periods will get impacted. Going too low on fuel can lead to a loss of periods altogether.
- We have to actively move away from diet culture and toxic beauty standards that tell us we're not enough. Instead, we have to focus on nourishing our bodies, minds, and souls.
- Weight stigma is rife in the medical system, while disordered eating and body dissatisfaction are prevalent in our populations. Nutrition is about health, not weight. We've got to find better ways to support people.
- Focus on balanced nutrition with meals that have carbs, protein, and fat throughout the day. Increase the diversity in your food, incorporating lots of plant foods like fruits, vegetables, nuts and seeds, whole grains, and legumes.
- You have to eat multiple times per day. Learning the basics of cooking, having enjoyable snacks, stocking your freezer and pantry, prepping some meals, and perhaps recruiting professional help are all ways to move the needle on your nutrition.

- Water should be the drink of choice. Minimize or avoid alcohol as it can contribute to PMS symptoms and long-term health risks like heart disease and cancer.
- Work on your digestive function by eating in a relaxed environment, getting plentiful fibre and exercise, chewing your food, and perhaps speaking to your doctor if your bowels need support.
- It's best to consult with a naturopathic doctor or another qualified practitioner about your supplements, especially if you are on many, have a health condition, and/or are on medications to ensure that they're appropriate for you.
- The best way to support your menstrual cycle through diet and nutrition is to focus on the basics.

CHAPTER 30

Mindset and Lifestyle

Improving your health does not mean that you need a twenty-step morning routine or be part of the 4 AM Club (As a night owl, that sounds pretty terrible to me; if that works for you, then you do you!).

I'm not a behaviour change expert or a psychologist, but I've had to learn a lot about helping people shift their behaviours (including myself). After much trial and much error, what I can safely say is that there's no "perfect" way to implement change. I can also say that practice makes you better; moreover, failing, making mistakes, wanting to quit, questioning why you started in the first place, and not being great when you start a new activity are parts of the learning process. There might be some things that are effortless, but others not so much.

Lifestyle and mindset changes, in theory, might sound simple; however, altering habits and shifting our inner narratives are no easy feats. They are critical for personal growth, health promotion, and disease management.

We already know that health is constantly being informed by the changing realities of our lives. All of this rests on myriad factors that overlap, contradict, and are basically impossible to tie together in a neat little package (or sentence for that matter!).

What this means is that working on health can look like different things for each one of us. It can also look different for you based on which season of life you're in and what your priorities are.

Go To Sleep

Sleep is paramount for healthy menstrual cycles and overall health. Lack of (restful) sleep will trickle down to every part of your health including your sex hormones, metabolism, hunger cues, mood and energy levels, pain tolerance, and your ability to think and work. Lack of sleep is also associated with an increased risk of all-cause mortality.

I'm pretty sure you can think back to a time when you didn't feel your best because you didn't get proper shut-eye. Our decision-making skills take a hit when we don't sleep well.

Many health habits are tied to us feeling rested and not tired–everything feels like a chore when sleepiness and fatigue kick in.

Here are some general sleep tips for you:

- Implement a wind-down routine like dimming the lights, washing up and massaging your face, putting on some white noise or music, meditating or practicing deep breathing, drinking a cup of tea, reading a book, or a combination of a few different activities.
- Sleep in complete darkness. Melatonin, our sleep hormone that is also a potent antioxidant, is released by the pineal gland in the dark.
- Expose your eyes and body to daylight, especially in the morning. This has a positive impact on your sleep-wake cycle (i.e., your circadian rhythm), which affects your mood and everything else. Light therapy lamps can be helpful if you cannot access outdoor light for whatever reason, like during the winter or if you live in a basement.
- If you tend to wake up in the middle of the night, a protein and fat snack like a tablespoon of pumpkin seeds or some nuts before bed can be of benefit. This strategy may be of specific use to those with blood sugar dysregulation, who are undereating or restricting fuel, and/or when there is a big lag between the last meal of the day and bedtime.

- Temperature control is a point of contention in many households if you have other people in the house and/or in your bed. I know my husband and I have different preferences for temperature, degree of clothing, and thickness of blanket for optimal sleep (plus, we often have a doggy or two in our midst!). Figure out a system that works for you so you're not feeling too cold or too hot at night.

- If you have restless leg syndrome keeping you up, talk to a healthcare provider since nutritional deficiencies or electrolyte imbalances can contribute to that.

- Herbs like chamomile, passionflower, and catnip can help relax the nervous system and offer gentle sleep support. A warm cup of tea is a great way to wind down. Herbs can also be taken at therapeutic doses, but it's always best to consult with a knowledgeable practitioner before taking those.

- Meditation, mindfulness, and slow, deep breathing at a rate of five to six breaths per minute can help reduce the time it takes to fall asleep.

- If you're delaying sleep because you're scrolling through social media or trying to have some solo time, put a time limit so that it's not interfering with catching some shut-eye. I know as well as anyone that time tends to slip away when you're alternating between doom-scrolling and watching cute puppy videos in the middle of the night! Blue light from screens may disrupt melatonin production as well.

- Naps can be conducive to catching up on sleep to help you feel better (you might need to figure out nap timing so that you're not waking up groggy or disrupting night-time sleep).

- During perimenopausal years, hormone changes, frequent urination, and night sweats can interfere with sleep, so getting support for those can help you get restful sleep. There are a few tools that can be implemented here including herbs and hormone therapy.

- If you suspect a sleep disorder or sleep apnoea, a sleep study can be an important aspect of your assessment. Snoring, metabolic health changes, high red blood cell count on bloodwork for no apparent reason, clenching or disorder of the TMJ (i.e., temporomandibular joint, above the jaw), weight gain, unexplained fatigue, and/or not feeling rested despite sleep are reasons that you might want to get assessed for sleep apnoea.
- Acupuncture and massage therapy are beneficial for many health problems, and insomnia is one of them. If pain, jaw clenching, muscle tension, and/or stress are getting in the way of sleep, these therapies can be helpful for that as well.
- Counselling and therapy can be helpful, especially if stress or mental health is getting in the way of sleep.
- Taking an inventory of your caffeine and alcohol consumption may be a good idea if you struggle with sleep: decreasing the quantity and timing away from bedtime are prudent considerations.
- Regular exercise and spending time in nature benefit sleep quality. My best sleep happens after long days of physical activity in nature.
- Consider a natural or pharmaceutical sleep aid. There are numerous options to discuss with your healthcare team like melatonin, magnesium, GABA, herbs, combination products, and medications.

Rest & Recover

Rest, recovery, relaxation, and enjoyment are necessities for a full life and healthy ageing. These are the facets of life that complete your physiological stress cycle, are imperative to your resilience, and help you feel like you.

Grind culture and capitalism have led to so many of us feeling like we don't have the time or capacity to rest and recover. It has also led us to believe that we need to *deserve* rest, constantly be doing things, sacrifice sleep, and put ourselves last.

Then, there's the guilt that creeps up when you do, in fact, try to rest or ask for help. I can't say it better than the Nap Bishop, Tricia Hersey: "Stop saying rest is a luxury or a privilege. It is not–it's a human right."[32] For more on rest, I highly recommend her book, *Rest is Resistance.*

Our screens and social media have offered us a lot of entertainment, connection, representation, inspiration, and information. Concomitantly, they have stolen away time to engage in pleasurable activities, connect deeply with ourselves and others, play, nap, and be present. Our ability to stay focused and off our phones is decreasing while our attention spans and ability to retain information are shifting.

Disconnecting from the online space, from our work, and our to-do lists might just help us by providing space for creativity and imagination, less comparing to others, and more living of our own lives.

Another aspect of rest is self-care. Self-care activities are the things that bring us pleasure, peace, and joy. They can be anything you want them to be and don't need to line up with other people's self-care activities. They can be done solo, indoors, outdoors, with loved ones like family, friends, pets, etc. There are no set rules. They can and will probably change over time.

For me, this can take the form of nature walks, reading a book, paddleboarding, cuddling with the dogs, spending time with family and friends, crafting cards, playing video games, travelling to explore new places, and playing saxophone with my community band. Sometimes, preparing a meal is self-care, while it's a chore at other times. Ultimately, self-care is personal to you. There's no set time limit, and it is not the things that you *need* to function like showering or sleeping.

---

[32] *Hersey. Rest is Resistance, 2022.*

BOX 27. The magic of nature

Fresh air, green spaces, and sunshine have all been shown to support health. *Shinrin yoku*, a concept that emerged from Japan meaning forest bathing, has been shown to improve anxiety, depression, fatigue, irritability, cortisol levels, blood pressure, and confusion scores in different populations, including young women. Nature also stirs up awe, an emotion linked to feeling contentment, gratitude, and connection.

There aren't any specific studies on forest bathing and PMS as far as I've seen, but it's not much of a stretch to say that it can help with PMS symptoms given the positive impacts on health. I find there's no remedy for the soul like time spent outdoors, and I walk almost daily–rain/snow or shine–to get my nature time!

There are lots of ways to incorporate nature into your life. This can include sitting by a window with a plant or going out to the patio, backyard, porch, or park.

The principle of a few minutes being better than zero minutes applies here as well. I recommend outdoor time to patients regularly. In 2022, Canada hopped on the nature prescriptions train by providing practitioners in certain provinces (including where I live in Ontario) the ability to prescribe free Parks Canada passes to patients. And, as a big advocate, I jumped on board immediately.

Being outdoors and practicing restorative nothing are ways I practice mindfulness, paying attention to the present on purpose in a non-judgmental way.

Restorative nothing is exactly what it sounds like it's doing nothing intentionally to relax and fuel your mind and body. I sometimes do this by lying on the floor with a soft gaze or with my eyes closed, and sometimes when I sit in silence looking out a window. You, too, can practice restorative nothing. You can also practice mindfulness. This may be done in many of your daily activities and routines, like while exercising, listening to music, enjoying a meal, breathing deeply, taking a shower, praying, and journalling. Journalling is another tool that can support your mental-emotional health, especially for those living with chronic health conditions.

Mindfulness and meditation (and mindful meditation) have gained popularity in research and in the routines of people all over the world in the last couple of decades. The terms have become quite elastic to refer to a lot of things, with a lot of appropriation of these practices.

For your menstrual health, I would like to lean into the theme of slowing down, being in the present, and, perhaps, noticing what's happening on the inside and outside. This has an immense benefit on anxiety, depression, pain, sleep, and overall well-being. We always have so much on our minds. Making time to simply be is a step toward connection to yourself and your environment.

Compassion is Cool

Practicing compassion, kindness, and gratitude not only toward others but also toward oneself has also been shown to help in all areas of health, especially mental-emotional health.

These practices have been shown to improve mood, stress, muscle tension, sleep, and overall well-being. They allow you to step back and shift your perspective. You can engage in these anywhere, anytime, out loud, quietly, in a journal, or with other people for the tiniest of things to the most life-changing of moments.

We tend to excel at extending compassion and kindness to others ... not so much to ourselves, though. We are our own harshest critics. I don't know about you, but the voice in my head can get pretty darn nasty at times. It can be a voice that's difficult to quiet down.

You are not alone if you have negative thoughts or even ones of hate toward yourself. I've had the privilege and honour of being the caretaker of the thoughts and stories of thousands of people over my time thus far as a naturopathic doctor, birth doula, and human on this earth. Not everyone gets this chance to see that many people up close and personal physically and emotionally, glimpsing even the dark recesses of the mind.

I have seen unfiltered real people of many ages, sizes, shapes, backgrounds, and health statuses in close proximity at different stages in life. I am more aware that the thoughts I have are shared by people around me. I have a deeper understanding of how bodies change, values change, priorities change, and life changes. I have gained insight into the biases I hold. I am learning to appreciate my body and accept the parts of my whole self I thought were "imperfections" (but are simply part of the diverse and shared human experience).

More and more of this content is also showing up in the media and online, with people sharing real and raw stories, unfiltered and unposed photos pictures and videos of themselves in different lighting, and baring their souls. While most of what we see in the media and online is curated and edited, authenticity and real talk are also becoming more commonplace.

Body dissatisfaction is running rampant in society. This is despite how our bodies are so cool, can do so many things, and serve us in a multitude of ways. We've been marketed to our whole lives. We've been told that we're not enough and we're the root of all our problems. We spend so much time meticulously picking ourselves apart that it consumes us, robbing us of our precious time, sleep, resources, and energy. We've already discussed diet culture, patriarchy, white supremacy, and capitalism, so I won't go on about how these systems divide us and tell us who's worthy and who's not.

We need to shift the narrative and have more compassion and gratitude for our bodies and our whole selves. This may require the safety of friends or family, a therapist's support, a community-based setting, some solo time, affirmations, books and podcasts, or a combination approach.

The thoughts you think and the words you speak have an impact on your health. Your subconscious mind is always listening to how you talk to yourself and even starts to believe what you're saying even when it's not true. Pay attention. Mindset, stress, trauma, and undealt-with emotions can all leave marks on your physical health.

You don't have to love your periods; however, appreciation for what your body can do and constantly does for you, and perhaps coming from a more neutral (vs. charged) standpoint might be a good place to start. As discussed in Chapter 24, your perception of your menstrual cycle and periods can impact your experience of this cycle. Here are some neutral affirmations and questions to help you along:

- I'm not supposed to feel the same way day in and day out
- What can I do for myself right now?
- I am more than my physical body. I am more than my thoughts.
- My hormones change and they affect how I feel

Connection is Key

In today's world, it's easy to get caught up in the weeds of daily life, forgetting to spend time and foster our relationships.

Connecting with others, whether in a small way or extremely meaningful way, is crucial to our well-being. We all need social connection–it forms an essential part of our lives.

Putting time and effort into our relationships and communities influences how we relate to ourselves, others, and our lives. The loss of interpersonal engagements–from surface-level interactions at the grocery store to meaningful heart-to-hearts with friends–during the COVID-19 pandemic lockdowns negatively impacted the health and well-being of people of all ages; cognitive health, mental health, heart health, and overall life satisfaction all suffer when we're cut off from others.

I encourage you to prioritize the individuals and communities that foster a sense of acceptance, belonging, joy, friendship, love, care, and the other things you want to feel and be part of.

Communities like a group of friends, a family circle, a business mastermind, a team, or an online forum revolving around a commonality of some kind allow us to be part of something bigger than just ourselves. We all know what it's like to feel understood, validated, inspired, and supported. Being part of a community offers a sense of belonging and enjoyment. It also helps normalize the human experience.

I, like you, straddle many identities based on my background, upbringing, values, geography, profession, family, aspirations, political views, and more. Being a part of different ensembles and communities, including my family and friend circles, allows me to stay connected to those parts of myself.

I know I have gained so much more than knowledge from being part of different groups and communities over my life–from sports teams and concert bands to non-profit boards of directors and professional masterminds. I've discovered, rediscovered, and even shed parts that no longer serve me through these different groups. I would not be the person I am if I didn't have these people and experiences. They have enriched my life and broadened my horizons, my dreams, and my goals. They've made me understand more deeply than before that it's okay to stay the course and it's also okay to make a turn; sometimes letting things go allows for new things to grow; just because I can doesn't mean I should; everyone has evolving goals and priorities; and, we're all doing our best.

Move Regularly

I think we all have some idea of the benefits of exercise. Unfortunately, this doesn't necessarily translate to us exercising. I know it can be a difficult muscle to build, pun intended!

Movement is crucial for mobility, strength, metabolism, bones, pain, longevity, healthy ageing, elimination of waste, healthy hearts, mood and mental health, and so much more. Exercise and movement are evidence-based interventions we have for menstrual health, including PMS and period pain.

Non-exercise activity is an area that is underrated but extremely important for health. These are the kinds of movements that are a part of daily life that are not formal exercise, like going down to the kitchen from an upstairs bedroom, cleaning or cooking, a leisurely walk, etc.

I've been discussing the impact of this a fair bit more with patients in recent years, especially since the COVID-19 lockdowns that took away people's non-exercise activities like trips to the grocery store, nights out with friends and family, and the time milling about the house trying to get ready to go out for the day to school, work, or other routine activities.

We know that our sedentary lifestyles contribute to increased chronic illness risk, especially cardiovascular disease, bone loss, and cognitive decline. Non-exercise activity isn't a replacement for formal exercise, but adding more is very helpful. When it comes to movement, two minutes is better than zero: walk, use the stairs, stand up more, stretch, use the bathroom farther away, put on music and dance, play with your pets or kids, pump your fists in the air, do something!

The same principle of two minutes is better than zero applies to formal exercise. For menstruators, bone health, cognitive changes, and cardiovascular risks all change once we approach menopause and especially post-menopause.

Exercise, both resistance and cardio, protects against these risks. The best time to get started is while you still have higher amounts of estrogen during the cycling years. Exercise is also one of the evidence-based recommendations we have to support people in the menopause transition for symptom management.

The Canadian guidelines for exercise in adults to lower the risk of heart disease, stroke, premature death, certain cancers, osteoporosis, high blood pressure, mood, and diabetes are as follows:[33]

- 150 minutes per week *minimum* of moderate-vigorous levels of activity
    - Moderate intensity refers to activities where you work up a bit of a sweat and your heart rate starts to rise like a brisk walk or a light jog

---

[33] *Canadian Society for Exercise Physiology, Canadian Physical Activity Guidelines Canadian Sedentary Behaviour Guidelines, 2012*

    - Vigorous intensity or high intensity refers to activities where your heart rate rises significantly, and you can feel out of breath like running or biking uphill
- Each bout of exercise should be at least ten minutes long
- At least two days of strength or muscle-building exercise targeting your big muscle groups is beneficial
- Working on different aspects of the physical body like cardiovascular training, building and preserving muscle mass, flexibility, and balance is encouraged

To meet these requirements, your exercise routine can look like around twenty minutes of exercise daily. It could also look like thirty minutes of exercise five days per week. These are minimum requirements, so I encourage you to do more (as do the guidelines!).

There are so many different ways to incorporate exercise: jog, swim, dance, use resistance bands, use free weights, join a team, get on the elliptical, do bodyweight exercises, climb, go up and down stairs, hop on the trampoline, take a class at a gym or community centre, row, hire a personal trainer, follow along with YouTube videos, and I could go on! Pick something you enjoy and/or something that you'll actually do, and make sure to challenge yourself since that's how you'll get stronger, fitter, faster, and better.

If this is too much to get started with, that's okay! Start somewhere and move from there. Anything is better than nothing even if it's well below the recommended minimum. Guidance from a fitness professional, especially as you're learning the ropes or trying to level up, will expedite your journey while minimizing the risk of injury.

BOX 28. Do I need to worry about high-intensity exercise and cortisol?

Your heart rate might spike a bit more and you might get winded more easily as you get started with moderate- and high-intensity exercise. This is natural. You'll get better with time as the body adapts. In hormone and exercise conversations online, there is much confusion surrounding high-intensity exercise and cortisol, our stress hormone.

Cortisol spiking when you add physiological and physical stress with exercise, especially intense activity, is part and parcel of exercise physiology and the adaptation response. Don't fear this! It's how the body works. Regular exercise improves cortisol profiles over time. Plus, high-intensity exercise is particularly helpful for cognitive and heart health.

Sure, you could be overdoing it (for e.g., doing high-intensity exercises six times per week for hours on end) which is a recipe for poor recovery and feeling unwell; however, more often than not, it's food restriction and a lack of rest days while engaging in intense exercise that makes feel people unwell after exercise, especially when we consider that many people use food restriction and high-intensity exercise to shrink their bodies.

So, I encourage you to get started somewhere. We have too many people under-exercising. We need to simplify exercise and remove barriers to movement.

To improve cardiovascular capacity, muscle growth, and recovery, you need consistent and progressive training, adequate fuel, and proper rest. Consider getting professional support.

We have to learn to ditch the all-or-nothing thinking. Instead of foregoing exercise completely if you can't get a full thirty-minute workout in, exercise for ten minutes. You can do another ten minutes later. Even if you don't, you still did ten minutes! If you have kids in your home, modeling healthy habits helps to foster those habits in them as well.

Let the definition of consistency change over time: perhaps consistency is a minimum of thirty minutes of exercise over the week regardless of how or on which days you achieve it. Then, it can change to sixty minutes over two days of the week. Then, it can move up to ninety minutes of gym time over the week, and so on. Done is always going to be better than whatever the perfect picture of exercising looks like in your head.

So, set a goal, put it into your schedule, and get started. You can reassess regularly and change things up accordingly. If you fall off the exercise bandwagon, simply hop back on–no one is judging you except for you. Remember, you've gotta do the exercise to get better at doing exercise–not being great at it, wanting to quit, it being difficult, and setbacks like injuries or loss are all normal parts of the process (and life!).

Most of my patients want to "age well" and "function optimally". When elaborating, this usually means they want to have their health, be of sound mind and body, and have the capacity to engage in hobbies and do the things they love while spending time with loved ones. Exercise is key to this physically, mentally, and physiologically. Aim for diversity in your movement that targets different aspects of health like strength, flexibility, cardiovascular conditioning, and balance! You can do hard things!

Endocrine Disruptors & Environmental Exposures

Endocrine-disrupting chemicals (EDCs) are compounds that can interfere with your hormones, and sometimes even mimic their action. They can be human-made or from natural sources influencing the body in numerous ways.

They can have acute implications on health, be stored in fatty tissue (accumulating over time), have additive effects, alter how genes are expressed, impact health at low-level exposures, and have ramifications for offspring and future generations.

The Industrial Revolution expedited the exposure to environmental pollutants that have irrevocably impacted not only the trajectory of human health but also that of all living things and the planet. Moreover, our exploitative practices have created astronomical changes in our ecosystems, which feeds back into all of this.

It has been difficult to gauge the exact cause-and-effect relationship of the many hormone disruptors out there. There are individual differences in how we respond to and eliminate these compounds. Scientific research has elucidated some themes that I'm sure will be expanded upon in the near future. We're likely looking at heterogeneous and cumulative effects over time.

Exposure to certain environmental toxins and endocrine disruptors has been associated with alterations in childhood development, early age at puberty and menarche, changes in metabolism, contributions to hormonal and menstrual disorders, immunological dysregulation and autoimmunity, generalized inflammation and allergies, fatigue, lung problems, cancer, and infertility.

Here are some of these compounds with their potential effects on the body:

- BPA (bisphenol A) found in receipts and (soft) plastics is carcinogenic and can bind to estrogen receptors strongly, stimulating estrogenic activity.
- Parabens in personal care products like shampoos and deodorants can mimic estrogen.
- Phthalates in plastics and fragrances can lead to cell death and contribute to thyroid issues.
- Dioxins in herbicides that accumulate in the fatty tissue of animals (meat, fish, and dairy) can interfere with embryonic development and certain cancers. These have also been implicated in endometriosis.

The theme of avoiding and limiting exposures as much as possible is being corroborated more and more through long-term and nascent research exploring these topics.

Decreasing your interactions with these compounds to the best of your ability is a great place to start. Unfortunately, these compounds are fairly ubiquitous: in personal care products, makeup, cleaning supplies, food packaging, the air, groundwater, and soil. This is another area where marginalized communities and those living in poverty carry a disproportionate load with a low level of access to supportive measures.

I know all of this can sound scary, and my goal isn't to instill fear or have you throw out all the things that you have. However, you can start to shift your lifestyle slowly over time. Here are some strategies recommended by environmental medicine experts and research to start lowering your exposure:

- Ventilate your home by opening doors and windows regularly. Indoor air has been recorded to be worse than outdoor air in research. You can consider using an air purifier, especially when the air quality is poor and opening windows are not feasible (e.g., due to wildfires and smog). And, ditch the air fresheners.
- Dust and wet-mop your home regularly, and leave your outdoor shoes for the outdoors. Dust and contaminants from the outside world have been shown to contain a lot of particulate matter that's not great for our health.
- Take a look at your personal care products for common hormone disruptors, and replace them as you run out when or if necessary and feasible with products that have less potential for interfering with hormones. This includes perfumes, body mists, and colognes. Checking out your daily staples is a good place to begin. The products you don't use super regularly probably have less of an impact, so maybe those don't matter as much.
- Decrease cleaning product exposure. Good ole vinegar and water can clean a lot of household surfaces. This way you can whip out your other, more chemical-laden, products infrequently for specific instances. Using a face mask and gloves while using strong cleaners is a good idea.
- Work on quitting smoking if you smoke cigarettes or cannabis. Cigarette smoke affects your health as well as those around you. THC (tetrahydrocannabinol), the active compounding in cannabis has been shown to downregulate the brain hormone GnRH (gonadotropin-releasing hormone) and impact fertility. Minimizing your exposure to these substances is always a good measure.
- Decrease saturated fat consumption to ensure your cells and liver function well. Moreover, many hormone disruptors and other environmental pollutants tend to concentrate in the fatty tissue of the animals you eat, including heavy metals that often accumulate in big fish.

- Try to support policies and local businesses that have sustainable and environmentally conscious practices when possible. These businesses are trying to change the landscape of how and what we consume. This will hopefully impact infrastructure to have stricter regulations and sanctions for the big corporations contributing significantly to pollution and forever chemicals that not only impact human health but also our planet as a whole. (A girl can dream!) You boycotting these big corporations (as much as you're able to) helps because money is the language Big Business speaks.
- Local and national environmental and endocrine organizations usually have some resources around hormone disruptors. There are also many environmental scientists and doctors doing great work in this field, some of whom you can follow on social media.

There's a balance to be struck between avoidance and practicality. Your body is resilient, and you have natural detoxification pathways built in through the liver, urinary tract, skin, and gut.

Support your detox processes and metabolism by drinking adequate water, exercising and sweating, pooping and urinating regularly, sleeping well, managing stress, limiting alcohol exposure, and giving your body raw materials like fibre, antioxidants, vitamins, and minerals regularly.

As you do that, try to decrease your daily exposure using some of the considerations listed in this chapter and by following recommendations by those who know about this subject more than myself.

CHAPTER HIGHLIGHTS

- There's no "perfect" way to implement change. Changing behaviour and creating habits takes time and effort, usually more so at the beginning. It will probably also include failure or sucking at something. Keep at it, you get better at something by actually doing that thing.
- Lack of sleep can affect sleep, mood, energy, pain tolerance, metabolism, hunger, and your ability to make decisions. If you struggle

with sleep, talk to your healthcare team. A relaxing tea, working on mental health, exercise, daylight exposure upon waking, acupuncture, supplements, and a wind-down routine can all be part of good sleep hygiene.

- Anything is better than nothing when it comes to exercise and moving in general. The guidelines for adults in Canada are to incorporate around 150 minutes of moderate to high-intensity activity spread over at least three days. A combination of cardio, resistance or strength training, mobility and flexibility exercises, and non-exercise activity throughout the week is recommended.
- Rest, recovery, relaxation, and enjoyment are necessities to feel well. Utilize your time to fill your cup, spend time in nature, connect with loved ones, and on activities that bring you peace, pleasure, and joy.
- Endocrine disruptors are compounds that can interfere with your hormones, and sometimes even mimic their action. Decreasing exposure and providing your body with the means and activities to process and eliminate waste products are good places to start.

CHAPTER 31

Advocating for Yourself and Your Cycle

Imagine a medical drama on TV: doctors are scrambling, collaborating, and poring over research to do everything in their power to understand what's going on with their patients ...

If only that were the case for more of us!

To be fair, most issues don't require that level of care, while some chronic concerns do receive that kind of care (but only in very specialized settings).

For most of us, though, this is pretty far from what healthcare offers. In fact, sometimes, championing for yourself is a heroic act of feminism. This is true even when your period problems or hormone-based concerns are getting in the way of your life.

When it comes to your general menstrual and hormonal health, routine testing like the kind laid out by policies and guidelines where you live is always going to be relevant.

In Canada, this includes things like your CBC (complete blood count), ferritin (iron stores), vitamin B12, glucose, cholesterol panel, HbA1c (three-month blood sugar analysis marker), TSH (thyroid-stimulating hormone), liver enzymes like ALT, urinalysis, cervical exams, and blood pressure monitoring. This allows you to track the basics over time and screen for the most common chronic issues. Then, your doctors should be layering measures according to your symptoms, age group, health history, family history, health goals, and experience in your body such as mammograms, colonoscopies, ultrasounds, CTs, MRIs, hormones, etc.

The conventional medical system, especially where I am in Ontario, Canada, does fairly well for acute and emergency care. However, the system is far from perfect, and even more so when it comes to chronic illness and hormonal health.

Even in the 2020s, menstruators face significant normalization of period problems, lack of validation, dismissal, gaslighting, and discrimination based on anything from race and gender to age and appearance. Short appointment times and one-concern-per-visit appointment structures are communication barriers and missed opportunities to educate patients.

Then, there are other factors such as disparities in quality of care, poor government funding for sexual and reproductive health education and programs, a lack of focus on the impact of social determinants of health such as income and working conditions, a low degree of consideration for someone's background and values, and long wait times to see specialists. In fact, finding a family doctor has been a huge challenge across Ontario in the last few years; millions of people currently do not have a family physician.

How medical care is delivered hasn't kept up with the care that people need, especially for complex and multifaceted non-communicable diseases. Preventive medicine and health promotion are not emphasized enough–although that's getting better with time–and reactive medicine is fairly standard. Don't get me wrong, there are situations where a wait-and-see approach is warranted; however, if you're feeling unwell and it's been a while since you've been waiting and seeing, then we need further action.

Pockets of our population, rightfully so, have a deep mistrust of medical intervention. With a history that comprises exclusion, stereotyping, eugenics, dismissal of traditional medicines, and experimentation on Indigenous and Black people, this should come as no surprise.

Bedside manner factors into how patients view their medical appointments as well; it significantly influences patient satisfaction. This is especially true for long-term doctor-patient relationships like that with a family doctor.

Depending on how you respond to others, the state you're in, and the way they communicate, a doctor's demeanour can affect how you feel about your care even if the treatment you received is the best option for your concern.

That said, there *is* a difference between a doctor who is direct and lacks tact, versus one who is condescending and dismissive. Unfortunately, some doctors gaslight, are simply not great at their jobs, choose their profession for the wrong reasons, are disinterested in helping patients, or have huge egos. This includes doctors who say things like, "*Go to a store, buy a fan, and ride your hot flashes out*" to a suffering perimenopausal patient (true story, unfortunately).

Thankfully, there are many great practitioners out there! However, a well-meaning doctor can still offer care that is incomplete, misguided, or culturally inappropriate. A caring doctor can also be going through personal struggles including burnout–a real problem health professionals face–and have biases, flaws, bad days, and other external and internal limitations.

As my practice and experience have grown, I've had to come to terms with my own biases, limitations, and knowledge gaps. I sometimes cringe at some of my articles and blogs floating around the internet as well as naturopathic care from my past that was well-intentioned but not entirely helpful, complete, practical, or efficacious. I've had and will continue to have seasons of steep learning curves. I can only continue to do my best as I learn, unlearn (I've needed to do a bunch of that!), keep up with evolving research, and move forward.

I see patients with different backgrounds and value systems: those who want more agency over their body and health, those who want to work on prevention and wellness, those who have great family doctors but want a collaborative approach, those who feel unsupported and dismissed, those who are trying to get a better understanding of their health and risks, those who are exploring their options, those who want a holistic approach, racialized patients who feel misunderstood by other practitioners or are simply looking for someone who they can relate to and vice versa, Urdu- or Hindi-speaking patients, and more.

The common theme always is that patients want someone to *really* listen to them and to have conversations *with* them. That is what patient-centred care is about.

Active listening and discussion are therapeutic in and of themselves; they also serve as tools to reach an accurate diagnosis and engage in shared decision-making. I have the luxury of time with patients as NDs in Ontario operate outside the public healthcare system.

We know that longer appointment durations than the five- to seven-minute appointment typical with a medical doctor in the province is the evidence-based way to offer care; however, it is difficult to achieve when the demand is high and the supply of doctors is low (amongst other infrastructural barriers).

There's an added layer of complexity with hormonal concerns; hormones are happening to us, and we are happening to our hormones. We also all have different sensitivities to the same levels of hormones. As we've learned throughout this book, you can't draw a straight line from a hormone level to how you're feeling, so having test results that are within normal limits doesn't offer the full picture.

Family doctors are often ill-equipped to deal with hormone-based diagnoses due to the breadth of conditions they see in practice and the constraints on their time. This is where a sexual and reproductive health specialist referral, like a gynaecologist, endocrinologist, or fertility clinic, comes in. (It's also where your allied healthcare team including a naturopathic doctor, pelvic physiotherapist, acupuncturist, etc. can come in!). While some family doctors are quick to refer to specialists, others need to be convinced, especially when the condition is multifaceted. And, of course, some do not listen at all.

Practitioners need to listen to their patients' experiences, embody the fact that they don't know everything and have biases, and be open to a multimodal approach. This is one of the reasons why gathering data on your body and tracking your progress over time is helpful: so that you can relay it to your healthcare team to elucidate what's going on.

That said, patients also need to understand that doctors have lots of training and experience, and probably got into this field because they wanted to help others. What this means is that it is part of their job to discern the salient information from less important bits. As such, not every morsel of information you've gathered is relevant to reaching a diagnosis or creating a treatment plan. I have held space for so many tears in my time as a practitioner. It can be emotional to talk about health, to reconcile with a diagnosis, and to make tough decisions.

There is a balance to be struck somewhere.

Systems of oppression and grind culture have robbed us of the opportunity, capacity, and right to trust our bodies, trust our intuition, and trust the connections we make. We have to get back in touch with our bodies, our people, and our planet with the understanding of how pervasive these social and structural constructs are in our personal lives, in modern medicine and healthcare, and the world at large.

On the flip side, we know that the connections we make and things we recall can be biased and inaccurate, so we have to be open-minded when testing our hypotheses and investigating what might be happening. Both, the patient and the doctor, need to approach care with curiosity, consideration, and collaboration.

Again, there's a balance to be struck.

Some healthcare providers are skilled at conducting assessments and asking questions that will allow them to make diagnoses, requisition the appropriate tests, put referrals in place, and/or implement an appropriate treatment plan. However, some doctors are not, which means that you need to share information without them asking. Your doctors don't know what they don't know.

I know that speaking to your healthcare team can be an uphill battle at times, but it is one worth fighting if you're not getting the care you need and deserve.

Here are some tips and considerations to help make your healthcare appointments more productive:

- Track and chart your cycle, especially your symptoms of concern. It's the single most effective tool you have to understand what's happening with your hormonal health, track your progress over time, connect the dots, and communicate with your healthcare team clearly and specifically. We also forget things and memories can change so this is a good way to remember some of the details that get fuzzy over time.
- When you do have a specific concern, make quantitative and qualitative notes on it. Be specific and succinct in your appointments and use the data you gathered while tracking. Here are some considerations that can be summarized in a few sentences as you share what your symptoms and experiences are:
    - Intensity: how severe is your issue?
    - Frequency: how often is it occurring?
    - Duration: how long has this been happening?
    - Change and progress: how is it shifting over time?
    - History: have you had this before?
    - Family history: do you have a family history that is related to this? If you're unsure and can, reach out to family members. For example, GPs don't typically assess for Hashimoto's thyroiditis, i.e., they do not run antibodies against the thyroid. However, if you have a family member with Hashimoto's or another autoimmune condition, then the likelihood of them running those tests is higher. Even if you think a condition in your family is irrelevant, let your doctor know.
    - What have you tried (and for how long)? What has worked? What hasn't? Write down all the strategies you use to manage your symptoms.
    - Provide context as to how your condition impacts you and your quality of life. This is important to communicate because not everyone has the same goals and desires, and this

    can tip the scale as to how your doctor approaches your case. Tell your doctor that you have to miss work or school, that your relationship with your family suffers, that your mood and emotions are difficult to handle, etc. That context can help your doctor shift their mindset around your care as well.
    - Ask them what testing can be done, what your treatment options are, and what happens if you take no action

- In Canada, the family doctor is the entry point to other specialists, so we need to be able to communicate clearly with our family doctor. If a referral isn't offered, then, unfortunately, oftentimes the onus is on us to ask them for referrals to gynaecologists, endocrinologists, psychiatrists, fertility clinics, dermatologists, etc.

- Speak up and use your voice. I am very familiar with the angst that comes with asking for what you need, but we've got to move through that discomfort and do it anyway. Practice in front of a mirror, with a friend or colleague, write things out so you have something to refer to and so you don't forget what you want to discuss, take someone with you if possible, or ask another care provider to write a note. Start with one thing, and perhaps something small. We've got to work that muscle for it to develop. It doesn't mean it'll get easier, but it will probably be easier to manage the discomfort that you feel in those situations.

- If your doctor has given you a diagnosis and/or treatment, you can ask them questions and request resources. You can also ask them for time to think about it. Here are some considerations when starting a medication or specific diet:

    - What are the pros and cons?
    - What happens if we do nothing in the short term and long term?
    - How long do I do this for?
    - What's the best-case scenario and the worst-case scenario?
    - How do we monitor progress?

    - What are the other options, and what else can I do that might help?

- If you're curious about whether a symptom is normal or not, ask your healthcare team. Instead of guessing and going down a Dr. Google rabbit hole, make a note so you don't forget to ask.

- Ask your doctor for a differential diagnosis list. This is essentially all the diagnoses that could explain what's going on with you. It is a doctor's job to go down a list of possibilities as part of your assessment. This is also how they determine what questions to ask you, which tests to run and what the treatment options are.

- If you're not feeling well or like yourself and your doctor tells you, "You're fine, your blood work is fine," then keep investigating, perhaps with another doctor if possible. A second opinion can be helpful. In a similar vein, if your doctor is hyperfocusing on an area of your health or life that is not relevant at the time, remind them why you're there. For example, if your doctor is focusing on your fertility or weight when you're actually consulting with them for fatigue, ask them how that's relevant and bring them back to why you're there.

- Gather a team of healthcare professionals who care about you and your well-being *and* will take into account your whole body and lifestyle into consideration. Sometimes this is not possible due to finances, accessibility, or other situations, but feeling supported in your healthcare is an important part of your health. There is a huge variation in the kind of support different doctors offer.

- Doctors forget things (they are human after all!). Yes, they have chart notes, but hearing how important something is to you can help direct your care. Repeating yourself and offering relevant context can sometimes help you receive the healthcare you need. For example, if you have low energy, brittle nails, period changes, hair loss, and a family history of autoimmune thyroid issues, that family history piece can be the tipping point in you getting a full thyroid panel tested versus only one thyroid parameter tested.

- Make notes during your appointment so that you can refer to pertinent aspects of your assessment, discussion, and treatment plan. My patients leave with a written treatment plan because, as humans, we forget things. Not only that, I want to make sure that they understand why we're doing something, how long it'll take, and what to look out for. Even with lifestyle interventions like exercise and nutrition, I don't simply tell patients to go exercise and eat better. We come up with a more specific plan.

It may sound like there's a whole lot of work we need to do for our health (which is true!). Of course, it would be amazing if we had a better system with longer appointments in place and we had more doctors with better bedside manner and a more rounded understanding of the multifaceted nature of hormone health and humans in general.

While change continues to trickle into our healthcare system(s), I hope that this offers some food for thought as you navigate your health and healthcare so you can achieve your goals, feel better, and spend your time in all the ways you want to.

CHAPTER HIGHLIGHTS

- Routine testing and check-ups allow you to monitor your health over time, screen for common health conditions, and a chance for you to discuss any concerns with your doctor(s).
- Menstruators face significant normalization of period problems, lack of validation, dismissal, gaslighting, and discrimination in healthcare appointments. Systems of oppression infiltrate all aspects of our health and life: from how we feel about ourselves to the modern medical paradigm.
- You have to be the one who cares most about your health. This means that you need to advocate for better care.

- Tracking your cycle, making robust notes about areas of concern, asking for a differential diagnosis list, discussing the nature of your treatment strategy, asking for resources, taking time to mull things over, and writing down the things you want to discuss and your doctor shared are all ways you can make your appointments more effective.

CHAPTER 32

Where Do We Go From Here?

The average menstruator has over 400 periods in their lifetime. That is a heck of a lot of menstrual cycles to go through without learning about them properly.

Education is a basic human right; this includes sexual and reproductive health education. Curiosity, choice, and informed decision-making in your healthcare are also rights you have.

It's mind-boggling that menstruators learn so little about this cycle that affects our lives on the regular for approximately forty years. (And then we learn even less about what happens as menopause approaches and beyond!) Ultimately, we have to view menstruation from more than a biological lens; we have to consider sociopolitical, cultural, and economic influences.

Improving your health and well-being aren't finite or stationary goals; they are moving targets that you get to define. The way forward is building sustainable habits that come from a place of curiosity about your body and goals, understanding that bodies and priorities change, and that health and wellness are journeys that will require some degree of effort. If you have a chronic condition, you'll likely need to manage those conditions and your health more specifically than someone who does not.

Essentially, we're all going to be on different paths based on our genetics, privileges, geography, age, schooling, identity, finances, culture, support system, medical history, values, and more. That being said, we have common goals and we need to take care of each other and this planet we live on.

I hope this book serves as a regular resource in your life when you want to brush up on your menstrual cycle knowledge, support your menstrual and overall health, advocate for yourself in your doctor's office, and figure out what part you can play in the menstrual movement.

If you're looking for menstrual learning beyond this book, you can join the movement at www.periodliteracy.com where you can find no-cost and sliding scale-based programming on period health from menarche to menopause that will grow over time. There is also a list of resources in Appendix B.

Like most people, I'm very good at pointing problems out. Solutions and change are harder. I've tried, as much as possible, to not only point out some of the glaring deficiencies we have but to also give some direction to move forward while holding on to hope.

It can be easy to fall into cynicism and despair, especially when we're complicit in capitalist and oppressive models of living in some way, shape, or form. I know it can lead to cognitive dissonance as well. Every step toward collective liberation and dismantling these systems matters—from the smallest step to the largest leap—counts.

Advocacy work and making change require pacing, just like everything else in life. Make sure you're creating time and space for yourself as you try to make this world a better place.

Here are some considerations for you to support yourself and menstruators across the globe:

- Talk more about periods privately and publicly so we can continue to dispel menstrual stigma and taboo. Spread awareness through whatever means you're able to: educate, create music, paint, dance, speak, write, advocate, amplify people and groups, protest, etc.

- Ditch the menstrual euphemisms and say “periods” and “menstruation”. Use specific anatomical terms when relevant.
- Track your cycle and pay attention to the overarching trends. Practice listening to your body–it’s telling you things. You might get it wrong sometimes and might even need a grain of salt when interpreting what it’s saying, but that’s part of the learning process. Approach it all from a place of curiosity.
- Ditch the fads and stick to the stuff that's tried and tested (unless the fad is to eat balanced meals, move more, connect with loved ones, and do more things that light your soul on fire; if that's the case, then be trendy by all means!). Your health doesn’t hang on that “one” elusive thing; it is a product of genetics, age, upbringing, lived experience, culture, medication, geography, access to resources like healthcare and social support, income, privilege, education, and many other dynamic variables. Some factors will be modifiable; others, not so much.
- Explore what health and well-being mean to you. Work on the foundations of your health, like food, nutrition, sleep, mindset, fostering friendships and connections, self-care, exercise, pleasure, purpose, and rest so that you can be the best version of yourself, keeping in mind that what works for one person may not for another.
- Don’t settle for feeling unwell, advocate for yourself, and move toward your ever-evolving definition of health. Managing a complex condition is different from pursuing wellness and health promotion. Don’t dismiss an effective chronic illness management strategy, whether allopathic or naturopathic, because someone else has bad things to say about it. You get to decide the best course of action with the help of your medical team while weighing out risks and benefits. You’re also allowed to change your mind.
- We are all cyclical beings, and just like everything else in nature, we go through phases and change.

- Continue to grow your understanding of sexual and reproductive health as well as the multifaceted nature of menstrual health from an intersectional lens. Engage in conversation, be curious, debate, and expose yourself to different points of view. Remember: to be human is to hold many truths all at once!
- Write to and meet with politicians to voice your concerns, and perhaps consider getting involved in policy-making. Politicians rely on their constituents to tell them what's important. Advocacy and awareness organizations often direct these kinds of campaigns. It's how we've seen funding for and changes in contraception care, screening mammograms, period product accessibility, and perinatal mental health support in the last few years in Canada. We desperately need more folk who will engage in these activities.
- Get involved in menstrual research, as a researcher, participant, or supporter.
- Support advocacy groups with your time, money, social platforms, allyship, creativity, and/or effort, while boycotting and divesting from perpetrators of injustice. Menstrual activism and health equity work are parts of dismantling systems of oppression. Every little bit helps.

My advocacy work has usually taken a quiet, behind-the-scenes route, as I've never felt like I had anything to say worth listening to. However, I know that's not true. Many of you probably feel similarly.

Please, use your voice and take up space in ways that feel right for you. To do that, you need to have your health, pace yourself, connect with others, and hold on to hope.

You deserve to feel well. There are too many menstruators who suffer their period problems without adequate care and support.

We need more menstruators in all kinds of roles, especially roles of leadership–from the personal and family level to the national and global level. We need people like you and me.

We need more people who care about the planet and its inhabitants, with the understanding that everything is connected and that the ecosystem of this planet relies on all of its sum parts.

You don't have to do it all, but all of us can do something. That's the way we build a world that is more equitable, inclusive, and compassionate.

ACKNOWLEDGMENTS

This book has been a labour of love (and sweat, tears, and years!). What started out as a simple book on menstrual physiology took on a life of its own and morphed into much more.

My name is on the cover, but it would not exist without generations of knowledge, tradition, advocacy, suffering, activism, research, experimentation, care, and discourse. It also would not be here in this form without the love and support of the people in my life.

My biggest thanks go to my husband and best friend, Dylan. From reading my drafts and telling me I have a problem with run-on sentences (most of which are shortened now), to making lattes and attempting to generate AI-derived book covers–all of which I rejected–you were vital to me accomplishing this feat (how's this for a long sentence?!). Thank you for celebrating my progress along the way, helping me word (and re-word) my thoughts, and for your generous encouragement and insights. I love you.

Next up is my cousin, Sarah (aka Saira). Your illustrations are exactly what I had hoped for and more. You have my heartfelt gratitude for all the back-and-forths, check-ins and catch-ups, cheerleading, and unwavering support for everything I do.

Thank you to my family without whom I wouldn't be the person I am. Ammi and Abbu, I'm grateful for the life and opportunities I've had, especially the experience of living in two very different countries–something that continues to shape the person I am growing into. Thank you for always supporting me and making me feel loved, even when we haven't seen eye-to-eye, for all your excitement about this book, and for all the snacks and food to fuel my writing.

To my other parents, Mum and Dad, thank you for always taking care of me especially in these last couple of years, engaging in conversations about the world, stocking up on my favourite snacks, supporting me and my work, and showing me, by example, how to go with the flow. Thank you to Ahm, for all the technical arguments on grammar and trivia, unconditional support, (unintended) life lessons that have helped shape my perspective in life, and for editing my blurb along with Niki. Thanks to Jess, who offered many resources and an amazing platform that helped me along my writing journey. And, thank you to Frankie and Danny, who can't read this (because they're dogs), but who sat with me as I wrote this book and offered much-welcome cuddles, walks, and comfort throughout this process.

Thank you to my editors who helped me polish my book. Clara Abigail, you gave me valuable feedback on the entirety of my book, especially the final chapter, and taught me that three sentences are better than one gargantuan sentence! Susan Keilor, your engaging comments, generous feedback, and guidance on when to use a hyphen after an adverb were extremely helpful (and validating!). Aleeza Arif, your thorough feedback and direction for the Introduction and blurb were exactly what I was looking for.

Thank you to my cover designer, Katerina Naskovski, who took all my ideas and notes and came up with something I wasn't expecting (in a very good way!), and to my book interior designer, Saqib Arshad, who helped make the inside look as good as it does.

Thank you to all my friends who are always generous with their encouragement and love. Special shout-outs to journalist extraordinaire, Sneh, for all your edits, feedback, and support for all the things in life since 2001; Olivia for holding me accountable and telling me I'm good enough when imposter syndrome struck (which was often); Manpreet, for suggesting that I break down the book into more digestible sections, musings on life and entrepreneurship, check-ins, and the shenanigans since our teenage years; Daina and Leenah, for reading parts of my draft and for your comments; Pari, for your edits on the blurb late on a Saturday night, cheerleading in life, and the insightful conversations about

SRH (inequities) over the years; Luc, Christina, and Emy, for encouraging me and reminding me to celebrate myself and this accomplishment along the way.

Thank you to my colleagues, mentors, support systems, and online community without whom I wouldn't have finished (or started!) my book when I did. Special thanks to Meghan and The Clinician Code for planting the seeds to take action; SPS for guiding me at the start; Jordan and The Confident Clinician Club for helping me hone my clinical skills and feel less alone, all the while offering some of the language in this book; my local entrepreneur friends, mastermind groups, colleagues, relatives, The Periodical email club, and Instagram community for your kind messages, poll answers, validation, encouragement, and suggestions.

A big thank you goes out to my patients; I wrote this book with you in my mind and heart. I have learnt and continue to learn so much from you and with you. Your vulnerability and experiences deeply affect me and my relationship with everyone and everything. I continue to learn the science and art of naturopathic medicine through you. Thank you for trusting me. I promise to continue to grow.

Thank you to all the activists, scholars, and advocates working toward closing the gender gap, taking up the cause of menstrual equity, instigating systemic change, championing racialized and oppressed populations, promoting planetary wellness, and trying to move us toward a more inclusive and equitable world. Thank you for your efforts and for leading by example.

And, last, but certainly not least, thank you, reader or listener, for your time and attention. I hope this book of mine offers something that you will carry forward with you.

ABOUT THE AUTHOR

Dr. Anne Hussain, ND is a naturopathic doctor and period literacy advocate based out of Canada. Her personal journey with polycystic ovarian syndrome (PCOS) and a lack of reproductive health education growing up in Karachi, Pakistan drives her passion for menstrual health education. Her private practice is located near Toronto, Ontario with a focus on hormone health, fertility, and PCOS. She obtained her Doctor of Naturopathy degree at the Canadian College of Naturopathic Medicine after completing her HBsc in Biochemistry at McMaster University.

Anne is the founder of The Period Literacy School, an online platform that offers free and paid programming for people with periods. She has appeared on Canadian national TV and has collaborated with organizations such as Naturopaths Without Borders, VegTO, and The Period Purse. She's been vegan for over seventeen years and loves to spend her time exploring nature, hanging out with family and friends, playing saxophone with her community band, paddleboarding, cuddling with her husband and two dogs, and learning about the human body.

Follow her on Instagram: @dr.annehussain
Website: www.annehussain.com
Period education website: www.periodliteracy.com
Author photograph: Tara Noelle
Cover design: Katarina Naskovski
Anatomical illustrations: Saira Hussain
Infographics: Anne Hussain

APPENDIX A

Glossary of Terms

Abnormal uterine bleeding (AUB): Blood loss from the uterus that is beyond the normal blood loss of menstruation, for e.g., extremely heavy menstrual bleeding, mid-cycle bleeding, bleeding after menopause, etc.

Absolute contraindication: A situation in which an intervention or treatment should not be administered as the risks outweigh the benefits.

Acupuncture: The practice of inserting thin needles into various parts of the body corresponding to different organ systems to promote the smooth flow of Qi and good health.

Adenomyosis: A condition with endometrium-like tissue found in the muscular layer of the uterus.

Allopregnanolone: Steroid hormone resulting from progesterone metabolism that is implicated in PMDD.

Amenorrhea: Absence of periods for three or more months.

Anaemia: A lack of oxygen-carrying red blood cells or haemoglobin protein which can lead to fatigue, hair loss, low mood, and other symptoms.

Anovulatory cycle: A cycle in which ovulation does not occur.

Assigned female at birth (AFAB): Assignment of sex before or at birth based on the external genitalia present.

Body Literacy: Understanding your body better to achieve wellness and identify any deficiencies or issues. This often includes an understanding of menstrual rhythms.

Cervical cancer: Uncontrolled cell growth (cancer) in the cervix of the uterus, typically related to infection with HPV (human papillomavirus).

Celiac disease: A gastrointestinal immune disease triggered by gluten consumption.

Cervix: The bottom part of the uterus that produces mucus and dilates during labour.

Cisheteropatriarchy: a sociopolitical system of power and control where cisgender, heterosexual males and the traits associated with masculinity are centred.

Coagulopathy: A condition that affects the process of blood clotting.

Cognitive behavioural therapy (CBT): Form of psychological intervention to help identify and change patterns of thought and behaviour.

Complete blood count (CBC): A blood test that assesses red and white blood cells.

Confirmation bias: Favouring and interpreting data in a way that supports one's existing beliefs, understanding, and values, typically ignoring information that is contrary to that.

Contraception: A means to prevent pregnancy.

Copper intrauterine device (IUD): T-shaped intrauterine device with copper coiling used for contraception.

Corpus albicans: Remnant of the corpus luteum after ovulation.

Corpus luteum: The empty follicle remnant that results from the release of an egg at ovulation.

Cortisol: A hormone produced by the adrenal glands as part of the physiological stress response.

Cyclical living: Living in sync with the ebb and flow of the menstrual cycle, understanding that the way one feels shifts slightly in different parts of the cycle.

Cycle tracking: Noting down changes experienced over the span of two or more menstrual cycles.

Decidualization: Changes in the endometrial lining of the uterus that start after ovulation and proceed into the early stages of pregnancy.

Depo-Provera: Hormone therapy injection commonly used for contraception containing the progestin called medroxyprogesterone acetate.

Dehydroepiandrosterone (DHEA): Sex hormone made primarily by the adrenal gland that serves as a precursor to other hormones like testosterone.

Determinants of health: The non-medical factors that affect health, healthcare, and health outcomes like geography, literacy, financial stability, etc.

Dihydrotestosterone (DHT): Sex hormone that is made from testosterone that is associated with pattern hair loss.

Dysmenorrhea: Medical term for pain associated with menstruation.

Emergency contraception: A form of birth control used to prevent an unplanned pregnancy in instances where there was no birth control used, use of contraception was not perfect, or if there was a contraceptive method failure.

Endometriosis: An inflammatory condition where tissue similar to the endometrial lining grows outside of the uterus.

Endometrium (also: endometrial lining): The innermost layer of the uterus that grows during the menstrual cycle and is shed during your period.

Endometrial hyperplasia: Overgrowth of the innermost lining of the uterus that may be precancerous.

Endometrial cancer: Uncontrolled growth (cancer) of the innermost lining of the uterus.

Epigallocatechin gallate (EGCG): An antioxidant derived from green tea used for various health concerns involving uncontrolled cell growth like fibroids.

Estradiol (E2): The most prevalent and strongest of the three forms of the sex hormone, estrogen, produced by the body. It is also used as hormone therapy in menopause.

Estriol (E3): One of the three forms of the sex hormone, estrogen, produced in the body. It is a weak estrogen and can be used vaginally for vaginal dryness.

Estrogen: Sex hormone that is responsible for the growth of the endometrial lining during the menstrual cycle and is associated with secondary sex characteristics such as breasts amongst other roles. It is also used in various forms of hormone therapy. See estradiol, estriol, and estrone.

Estrone (E1): One of the three forms of the sex hormone, estrogen, produced in the body. It is a weak estrogen and can be converted into estradiol.

Ethinylestradiol: The most common synthetic estrogen used in combination with birth control products.

Fallopian tubes: *see oviducts.*

Ferritin: Iron-storage protein typically measured in blood that may occasionally be elevated due to inflammation in the body.

Fertile window: The time frame during which conception can occur based on sperm survival period and ovulation.

Fibroids: Growths of connective tissue and muscle cells that may lead to heavy bleeding and painful periods.

Folic acid: An essential vitamin (B9) needed for cell division, especially in early pregnancy to reduce the risk of neural tube and congenital defects.

Follicular cyst: Fluid-filled pocket of tissue resulting from a lack of ovulation.

Follicular Phase: The first part of the menstrual cycle between your period and ovulation during which your follicles and endometrium grow.

Follicle stimulating hormone (FSH): A hormone produced by the pituitary gland in the brain that causes follicles to grow during the menstrual cycle.

Gender dysphoria: Distress related to an incongruence between assigned sex and gender identity.

Glycodelin A: A protein that inhibits implantation.

Gonadotropin-releasing hormone (GnRH): Gonadotropin-releasing hormone produced by the hypothalamus in the brain that is responsible for the release of follicle-stimulating hormone and luteinizing hormone.

Hashimoto's disease: An autoimmune disorder of the thyroid gland whereby antibodies are produced against the thyroid.

Health: A state of complete physical, mental and social well-being and not merely the absence of disease or infirmity, according to the WHO.

Hirsutism: Coarse and/or dark hair growth in those assigned female at birth in a pattern akin to those assigned male at birth, for e.g., on the face along the jawline and chin.

Human chorionogonadotropin (hCG): A hormone produced by a fertilized egg when pregnancy has occurred.

Hymen: A thin layer of tissue at the opening of the vagina that may take a variety of different shapes.

Hyperpalatable: Foods lacking in nutrients that are often high in fat, sugar, salt, and additives that make them attractive to the taste buds but not satiating, leading to easy overconsumption and deleterious effects on health over time.

Hypothalamic-pituitary-ovarian (HPO) axis: The communication system between the brain and ovaries that is responsible for the menstrual cycle.

Hysterectomy: Surgical removal of the uterus.

Iatrogenic: Induced unintentionally by a medical intervention of some kind.

***In utero*:** While in the uterus, from Latin.

Infertility: According to the ASRM: A disease, condition, or status characterized by any of the following: (a) The inability to achieve a successful pregnancy based on a patient's medical, sexual, and reproductive history, age, physical findings, diagnostic testing, or any combination of those factors, (b) The need for medical intervention, including, but not limited to, the use of donor gametes or donor embryos to achieve a successful pregnancy either as an individual or with a partner, and/or (c) In patients having regular, unprotected intercourse and without any known etiology for either partner suggestive of impaired reproductive ability, evaluation should be initiated at 12 months when the female partner is under 35 years of age and at six months when the female partner is 35 years of age or older.

- **Primary Infertility:** Infertility with no history of a past pregnancy.
- **Secondary infertility:** Infertility where a past pregnancy has been achieved.

Inflammation: Part of the body's immune response mechanism to protect and repair the body.

Informed consent: The practice of making educated decisions on how you want to proceed with your care under the guidance of your healthcare team.

Inhibin: A protein that stops the production and release of follicle-stimulating hormone.

Insulin: A hormone produced by the pancreas that regulates blood sugar and facilitates the movement of blood sugar into cells for energy production.

Intrauterine contraceptive (IUC): An intrauterine contraceptive, any contraceptive measure that can be inserted into the uterus.

Intrauterine device (IUD): An intrauterine contraceptive, typically without hormones.

Intrauterine system (IUS): A kind of intrauterine device that releases hormones, typically levonorgestrel.

Irritable Bowel Syndrome (IBS): A gastrointestinal disorder characterized by increased visceral hypersensitivity, altered gut microbiome, abdominal discomfort and pain, and variable stool consistency and frequency.

Kisspeptin: Protein in the hypothalamus of the brain that is involved in gonadotropin-releasing hormone and thyroid-stimulating hormone release.

Lactobacillus (plural lactobacilli): The primary bacteria genus that lives in the vaginal tract

Laparoscopy: A surgical procedure that is minimally invasive conducted using small incisions.

Leiomyoma (plural leiomyomata): Medical term for fibroids.

Leukotrienes: Inflammatory mediators involved in menstruation that are produced by the immune system.

Leuprolide: A hormone medication that is similar to GnRH and suppresses estrogen and testosterone.

Luteal cyst: Benign fluid-filled sac that forms from the corpus luteum after ovulation.

Luteal phase: The second part of the menstrual cycle from ovulation to right before the next period.

Luteinizing hormone (LH): A luteinizing hormone that surges mid-cycle and causes ovulation to occur.

Levonorgestrel: Progestin typically used in hormone intrauterine systems for birth control.

Mefenamic acid: A non-steroidal anti-inflammatory drug for pain relief.

Menarche: The first period ever.

Menopause: The natural state reached after twelve months of having no period at the end of one's reproductive life cycle.

Menopause hormone therapy: Use of estrogen and progesterone (typically) for the signs and changes that occur as natural estrogen and progesterone in the body (begin to) wane.

Menstrual cup: A vessel to contain menstrual blood that is inserted into the vaginal canal and sits there via suction.

Menstrual disc: A vessel to contain menstrual blood that is inserted vaginally and sits at the base of the cervix at the vaginal fornix.

Menstrual cycle: A combination of the events and hormonal changes that happen in the uterus and ovaries.

Menstrual health: A state of complete physical, mental, and social well-being and not merely the absence of disease or infirmity, in relation to the menstrual cycle.

Menstrual equity: Accessibility, affordability, and safety of menstrual products, menstrual education, and menstrual healthcare.

Menstrual toxic shock syndrome (mTSS): *See TSS.*

Menstruation: The process of shedding the endometrial lining during a menstrual cycle.

Myo-inositol: A compound often used to improve insulin sensitivity, improve ovulation rates, support pregnancy, and decrease androgens in those with PCOS.

Myometrium: The middle muscular layer of the uterus.

N-acetyl cysteine (NAC): An antioxidant used for lung concerns as well as menstrual concerns like PCOS to decrease androgens and improve ovulation rates.

Naturopath, aka naturopathic doctor (ND): A regulated healthcare professional who practices naturopathic medicine after successful completion of education at an accredited institution and standardized examinations, and is governed by a regulatory body that requires active maintenance of their license.

Naturopathic medicine: A primary healthcare system that focuses on the whole person, prevention, health promotion, and treating individuals using a combination of modern medicine and traditional methods of healing.

Non-steroidal anti-inflammatory drug (NSAID): A class of drugs used for relieving pain and inflammation including ibuprofen.

Nuva ring: Hormonal birth control that is inserted into the vaginal canal.

Oligomenorrhea: Infrequent and/or irregular menstruation.

Ovarian cyst: Fluid-filled sacs that can form on the ovaries.

Ovarian cycle: The changes in the ovary including follicle growth, ovulation, and hormone secretion involved in reproduction.

Ovulation: The release of an egg from a follicle that occurs after a spike in LH.

Ovulatory cycle: A menstrual cycle where ovulation occurred.

PALM-COEIN: Acronym describing the causes of abnormal uterine bleeding: polyp, adenomyosis, leiomyoma, malignancy, coagulopathy, ovulatory dysfunction, endometrial, iatrogenic, not otherwise classified.

Patriarchy: Societal and governing systems in which men hold power and privilege at the expense of women and other people.

Pelvic floor: Group of muscles at the base of the pelvis that support the reproductive organs.

Perimenopause: The timeframe around menopause that can start a decade or so before the last natural period.

Perimetrium: The outermost layer of the uterus.

Period literacy: Having a basic understanding of the menstrual cycle, cultivating an awareness of your menstrual cycles and the changes that happen over its duration, and holding all of that in the context of your life and the world we live in.

Pituitary gland: The part of your brain responsible for producing a variety of different hormones including FSH, LH, and TSH.

Polycystic ovarian/ovary syndrome (PCOS): an endocrine disorder characterized by at least two out of three of: (1) infrequent or anovulatory menstrual cycles, (2) clinical or biochemical signs of elevated androgens, (3) polycystic ovaries (and/or elevated AMH in adults).

Polyp: Tissue growth attached to an inner surface of the body like the uterus. It is typically benign but may be pre-cancerous.

Post-menopause: The time after menopause has been reached.

Precocious puberty: When the physical and hormonal changes associated with sexual maturation are reached earlier than expected: before eight years of age in those assigned female at birth and nine years in those assigned male at birth.

Premenstrual dysphoric disorder (PMDD): A hormone-based condition characterized by significant distress and/or emotional lability in the luteal phase only.

Premenstrual exacerbation (PME): a worsening of a pre-existing health condition in the luteal phase after ovulation.

Premenstrual syndrome (PMS): A set of physical and emotional hormone-based symptoms in the seven to ten days leading up to a period after ovulation.

Prenatal vitamin: A multivitamin designed to supplement a pregnant person with essential nutrients for an embryo to grow.

Primary ovarian insufficiency (POI): A diagnosis given when menopause is reached before 40 years of age.

Progesterone: Sex hormone associated with the luteal phase of the menstrual cycle and maintenance of a pregnancy.

Progestin: Hormones that are not made by the body and are similar to progesterone that is made in the body.

Progestogen: Term referring to progesterone and progesterone-like compounds called progestins.

Prolactin: A hormone that leads to milk production.

Proliferative phase: Part of the uterine cycle after menstruation until ovulation when the endometrial lining is building up.

Prostaglandins: Inflammatory mediators that are involved in menstruation (and other processes) that are produced by the immune system.

Puberty: Physical and hormonal changes through which a youth matures into having the ability to reproduce.

Qi: A concept from Traditional Chinese Medicine loosely translated to your vital energy or life force.

Recall bias: error due to an error in remembering past events or omitting details while retelling said events.

Relative contraindication: a situation that could be risky and particular caution should be exercised to ensure that the benefits outweigh the risks.

Rescue dose: dose of medication needed for breakthrough pain.

Secretory phase: Part of the uterine cycle from ovulation time until menstruation.

Self-care: Practices and activities intentionally undertaken for the fulfilment of one's joy, peace, pleasure, and creativity.

Shinrin yoku: Translated as forest bathing from Japanese, the practice of spending time in nature.

Testosterone: Sex hormone typically associated with masculinity that forms a precursor to estrogen and is involved in numerous processes in the body.

Thyroid stimulating hormone (TSH): A hormone released by the pituitary gland in the brain that tells the thyroid to produce its own hormones called T4 and T3.

Toxic shock syndrome (TSS): A rare medical condition in which a toxin called TSST-1 typically produced by the bacterium, Staphylococcus aureus, causes fever, rash, low blood pressure, and other symptoms that require antibiotic treatment. When associated with menstruation, it is called menstrual TSS or mTSS.

Tranexamic acid: Medication to prevent blood loss that is employed in cases of heavy menstrual bleeding.

Tubal ligation: Birth control method in which the oviducts are permanently blocked through surgery.

Tumour shrinking decoction: A Traditional Chinese Medicine herbal formula consisting of fifteen herbs commonly used for growths such as uterine fibroids.

Ulipristal acetate (UPA): A form of prescription-based emergency contraception.

Ultra-processed foods (UPFs): Foods made entirely or mostly from substances extracted from foods, derived from food constituents, or synthesized in laboratories from food substrates or other organic sources according to the NOVA Food Classification System.

Uterine cycle: The changes in the endometrial lining of the uterus in response to estrogen and progesterone.

Uterus: The primary organ involved in menstruation comprised of three layers called the perimetrium, myometrium, and endometrium.

Vagina: Fibromuscular tube on the inside of the body that runs from the cervix to the external genitalia.

Vasectomy: A form of male birth control in which the vas deferens, the tube that carries sperm, is sealed or cut and tied on both sides through surgery.

Vulva: Structures of the external genitalia associated with those assigned female at birth including the labia minora, labia majora, mons pubis, parts of the clitoris, vestibule, and urethral opening.

Wellness: Functioning optimally within your current environment according to the National Wellness Institute.

APPENDIX B

Resources

Periods & Related Resources

The Endometriosis Network Canada: https://endometriosisnetwork.com/
International Association for Premenstrual Disorders: https://iapmd.org/
PCOS Awareness Association: https://www.pcosaa.org/
Periods Gone Public by Jennifer Weiss-Wolf
The Period Literacy School: https://www.periodliteracy.com/
The Period Repair Manual by Lara Briden, ND
Society of Obstetricians and Gynaecologists of Canada: https://www.sexandu.ca/contraception/
Wild Power by Alexandra Pope
YourPeriod.ca: https://www.yourperiod.ca/

Period Tracking Apps:

Euki
Menstruation Nation
Period Tracker created by Aliza Aufrichtig of The New York Times on Google Sheets: https://www.alizaaufrichtig.com/period-tracker

Period Advocacy & Activism Non-profit Organizations (Canadian)

Free Periods Canada: https://www.freeperiods.ca/
Moon Time Connections: https://truenorthaid.ca/moon-time-connections/
The Period Purse: https://www.theperiodpurse.com/

Perimenopause and Menopause

The Hormone Repair Manual by Lara Briden, ND

Menopause Chicks: https://www.menopausechicks.com/

The Menopause Society: www.menopause.org

Menopausing by Davina McCall with Dr. Naomi Potter

Menoverse: https://hellomenoverse.com/

Preconception, Pregnancy, & Fertility

Evidence Based Birth: https://evidencebasedbirth.com/

Expecting Better by Emily Oster

FACTS (Fertility Appreciation Collaborative to Teach the Science) About Fertility: https://www.factsaboutfertility.org/

Fertility Matters Canada: https://fertilitymatters.ca/

It Takes Two... And a Uterus by Dr. Sarah Zadek, ND

The Natural Pregnancy Book, Third Edition by Dr. Aviva Romm

PregnancyInfo.ca: https://www.pregnancyinfo.ca/

Taking Charge of Your Fertility by Toni Weschler

NOTES & REFERENCES

INTRODUCTION

1. Global Menstrual Collective. 2019. "Menstrual Health" coined by the Terminology Action Group. <www.globalmenstrualcollective.org>
2. Hennegan J, Winkler IT, Bobel C, Keiser D, Hampton J, Larsson G, Chandra-Mouli V, Plesons M, Mahon T. Menstrual health: a definition for policy, practice, and research.

CHAPTER 1. MENSTRUAL ANATOMY

1. Netter, F. H. 1. (2010). *Atlas of human anatomy.* 5th ed. Philadelphia, PA, Saunders/Elsevier.
2. Hoare BS, Khan YS. Anatomy, Abdomen and Pelvis, Female Internal Genitals. [Updated 2021 Jul 31]. In: StatPearls [Internet]. Treasure Island (FL): StatPearls Publishing; 2022 Jan-.
3. Nguyen JD, Duong H. Anatomy, Abdomen and Pelvis, Female External Genitalia. [Updated 2021 Jul 31]. In: StatPearls [Internet]. Treasure Island (FL): StatPearls Publishing; 2022 Jan-.
4. Puppo V. Embryology and anatomy of the vulva: the female orgasm and women's sexual health. Eur J Obstet Gynecol Reprod Biol. 2011 Jan;154(1):3-8.
5. Olson RM, García-Moreno C. Virginity testing: a systematic review. Reprod Health. 2017 May 18;14(1):61.
6. Grimes WR, Stratton M. Pelvic Floor Dysfunction. [Updated 2022 Jun 27]. In: StatPearls [Internet]. Treasure Island (FL): StatPearls Publishing; 2022 Jan-.

CHAPTER 2. MENSTRUAL CYCLE OVERVIEW & CHAPTER 3. PHASES OF THE MENSTRUAL CYCLE

1. Thiyagarajan DK, Basit H, Jeanmonod R. Physiology, Menstrual Cycle. [Updated 2021 Oct 30]. In: StatPearls [Internet]. Treasure Island (FL): StatPearls Publishing; 2022 Jan-.
2. Reed BG, Carr BR. The Normal Menstrual Cycle and the Control of Ovulation. [Updated 2018 Aug 5]. In: Feingold KR, Anawalt B, Boyce A, et al., editors. Endotext [Internet]. South Dartmouth (MA): MDText.com, Inc.; 2000-.
3. Teede HJ, Tay CT, Laven JJE, Dokras A, Moran LJ, Piltonen TT, Costello MF, Boivin J, Redman LM, Boyle JA, Norman RJ, Mousa A, Joham AE; International PCOS Network. Recommendations From the 2023 International Evidence-based Guideline for the Assessment and Management of Polycystic Ovary Syndrome. J Clin Endocrinol Metab. 2023 Aug 15.
4. Fukaura R, Ward A, Datta S. Delayed miscarriage inside an infected decidual cast: a rare complication of the Depo medroxyprogesterone acetate injection. BMJ Case Rep. 2021 Mar 15;14(3):e238583. doi: 10.1136/bcr-2020-238583.
5. Voliotis M, Li XF, De Burgh R, Lass G, Lightman SF, O'Byrne KT,Tsaneva-Atanasova K. The Origin of GnRH Pulse Generation: An Integrative Mathematical-Experimental Approach. Journal of Neuroscience 4 December 2019, 39 (49) 9738-9747.
6. Channing CP, Schaerf FW, Anderson LD, Tsafriri A. Ovarian follicular and luteal physiology. Int Rev Physiol. 1980;22:117-201.
7. Okada H, Tsuzuki T, Murata H. Decidualization of the human endometrium. Reprod Med Biol. 2018 Feb 1;17(3):220-227.
8. Yu JL, Su YF, Zhang C, Jin L, Lin XH, Chen LT, Huang HF, Wu YT. Tracking of menstrual cycles and prediction of the fertile window via measurements of basal body temperature and heart rate as well as machine-learning algorithms. Reprod Biol Endocrinol. 2022 Aug 13;20(1):118.

9. Fahs B. Sex During Menstruation: Race, Sexual Identity, and Women's Accounts of Pleasure and Disgust. 2020 Jul 25. In: Bobel C, Winkler IT, Fahs B, et al., editors. The Palgrave Handbook of Critical Menstruation Studies [Internet]. Singapore: Palgrave Macmillan; 2020. Chapter 69.

CHAPTER 4. MAIN MENSTRUAL HORMONES

1. Adams JM, Taylor AE, Schoenfeld DA, Crowley WF, Hall JE. The midcycle gonadotropin surge in normal women occurs in the face of an unchanging gonadotropin-releasing hormone pulse frequency. J Clin Endocrinol Metab. 1994 Sep;79(3):858-64.
2. Sadiq NM, Tadi P. Physiology, Pituitary Hormones. 2022 May 8. In: StatPearls [Internet]. Treasure Island (FL): StatPearls Publishing; 2022 Jan–.
3. Hamilton KJ, Hewitt SC, Arao Y, Korach KS. Estrogen Hormone Biology. Curr Top Dev Biol. 2017;125:109-146.
4. Delgado BJ, Lopez-Ojeda W. Estrogen. [Updated 2022 Jun 28]. In: StatPearls [Internet]. Treasure Island (FL): StatPearls Publishing; 2022 Jan-.
5. Kumar A, Banerjee A, Singh D, Thakur G, Kasarpalkar N, Gavali S, Gadkar S, Madan T, Mahale SD, Balasinor NH, Sachdeva G. Estradiol: A Steroid with Multiple Facets. Horm Metab Res. 2018 May;50(5):359-374.
6. Taraborrelli S. Physiology, production and action of progesterone. Acta Obstet Gynecol Scand. 2015 Nov;94 Suppl 161:8-16.
7. Tam A, Morrish D, Wadsworth S, Dorscheid D, Man SF, Sin DD. The role of female hormones on lung function in chronic lung diseases. BMC Womens Health. 2011 Jun 3;11:24.
8. Goletiani NV, Keith DR, Gorsky SJ. Progesterone: review of safety for clinical studies. Exp Clin Psychopharmacol. 2007 Oct;15(5):427-44.
9. Cable JK, Grider MH. Physiology, Progesterone. [Updated 2022 May 8]. In: StatPearls [Internet]. Treasure Island (FL): StatPearls Publishing; 2022 Jan-.
10. Okada H, Tsuzuki T, Murata H. Decidualization of the human endometrium. Reprod Med Biol. 2018 Feb 1;17(3):220-227.
11. Zhang S, Osumi H, Uchizawa A, Hamada H, Park I, Suzuki Y, Tanaka Y, Ishihara A, Yajima K, Seol J, Satoh M, Omi N, Tokuyama K. Changes in sleeping energy metabolism and thermoregulation during menstrual cycle. Physiol Rep. 2020 Jan;8(2):e14353.
12. Nassar GN, Leslie SW. Physiology, Testosterone. [Updated 2022 Jan 4]. In: StatPearls [Internet]. Treasure Island (FL): StatPearls Publishing; 2022 Jan-.
13. Lois K, Kassi E, Prokopiou M, et al. Adrenal Androgens and Aging. [Updated 2014 Jun 18]. In: Feingold KR, Anawalt B, Boyce A, et al., editors. Endotext [Internet]. South Dartmouth (MA): MDText.com, Inc.; 2000-.
14. Hossan T, Kundu S, Alam SS, Nagarajan S. Epigenetic Modifications Associated with the Pathogenesis of Type 2 Diabetes Mellitus. Endocr Metab Immune Disord Drug Targets. 2019;19(6):775-786.
15. Brent GA. Mechanisms of thyroid hormone action. J Clin Invest. 2012 Sep;122(9):3035-43.

CHAPTER 5. FROM MENARCHE TO MENOPAUSE

1. Klein DA, Emerick JE, Sylvester JE, Vogt KS. Disorders of Puberty: An Approach to Diagnosis and Management. Am Fam Physician. 2017 Nov 01;96(9):590-599.
2. Practice Committee of the American Society for Reproductive Medicine. Definition of infertility: a committee opinion. American Society for Reproductive Medicine. Washington, DC, 2023.
3. Schlievert PM, Davis CC. Device-Associated Menstrual Toxic Shock Syndrome. Clin Microb Rev. 2020; 33(3).
4. Rosner J, Samardzic T, Sarao MS. Physiology, Female Reproduction. [Updated 2022 Jul 7]. In: StatPearls [Internet]. Treasure Island (FL): StatPearls Publishing; 2022 Jan-.
5. Schwartz BI, Effron A, Bear B, Short VL, Eisenberg J, Felleman S, Kazak AE. Experiences with Menses in Transgender and Gender Nonbinary Adolescents. J Pediatr Adolesc Gynecol. 2022 Aug;35(4):450-456.

6. The North American Menopause Society. Menopause: the Journal of The North American Menopause Society. 2022; 29(7): 767-794.
7. The North American Menopause Society. Menopause 101: A Primer for the Perimenopausal. 2024. <https://www.menopause.org/for-women/menopauseflashes/menopause-symptoms-and-treatments/menopause-101-a-primer-for-the-perimenopausal>
8. Menopause: Siobán D. Harlow. Menstrual Cycle Changes as Women Approach the Final Menses: What Matters? Obstetrics and Gynecology Clinics of North America, Volume 45, Issue 4, 2018, Pages 599-611.

CHAPTER 6. THE IMPORTANCE OF CYCLE TRACKING & CHAPTER 7. THE INS AND OUTS OF CYCLE TRACKING

1. Duane M, Stanford JB, Porucznik CA, Vigil P. Fertility Awareness-Based Methods for Women's Health and Family Planning. Front Med (Lausanne). 2022 May 24;9:858977
2. Eschler, Jordan, Amanda Menking, Sarah E. Fox, and Uba Backonja. 2019. "Defining Menstrual Literacy with the Aim of Evaluating Mobile Menstrual Tracking Applications." *CIN: Computers, Informatics, Nursing* 37 (12): 638–646.
3. Fox S, Epstein DA. Chapter 54 Monitoring Menses: Design-Based Investigations of Menstrual Tracking Applications [July 25, 2020]. In: The Palgrave Handbook of Critical Menstruation Studies by Bobel C, Winkler IT, Fahs B, et al. Palgrave Macmillan; 2020.

CHAPTER 8. CONTRACEPTION CONSIDERATIONS

1. Handelsman DJ. Male Contraception. 2019 Jan 25. In: Feingold KR, Anawalt B, Boyce A, Chrousos G, de Herder WW, Dhatariya K, Dungan K, Hershman JM, Hofland J, Kalra S, Kaltsas G, Koch C, Kopp P, Korbonits M, Kovacs CS, Kuohung W, Laferrère B, Levy M, McGee EA, McLachlan R, Morley JE, New M, Purnell J, Sahay R, Singer F, Sperling MA, Stratakis CA, Trence DL, Wilson DP, editors. Endotext [Internet]. South Dartmouth (MA): MDText.com, Inc.; 2000–.
2. Daniels K, Abma JC. Current contraceptive status among women aged 15–49: United States, 2017–2019. NCHS Data Brief, no 388. Hyattsville, MD: National Center for Health Statistics. 2020.
3. Freeland, Chrystia, Deputy Prime Minister of Canada. "Government highlights next step to universal access to free contraceptives." March 30, 2024. Toronto, ON. <https://deputypm.canada.ca/en/news/news-releases/2024/03/30/government-highlights-next-step-universal-access-free-contraceptives>
4. Health Canada. Universal Access to Contraception. February 29, 2024. < https://www.canada.ca/en/health-canada/news/2024/02/backgrounder-universal-access-to-contraception.html>

CHAPTER 9. THE PILL

1. Black A, Guilbert E, Costescu D, Wagnet M, Whelan AM, et al. Canadian Contraception Consensus. SOGC Clinical Practice Guideline 2015: 37(10): 936-938.
2. ACOG Practice Bulletin No. 110: noncontraceptive uses of hormonal contraceptives. Obstet Gynecol. 2010 Jan;115(1):206-218.
3. Powell A. Choosing the Right Oral Contraceptive Pill for Teens. Pediatr Clin North Am. 2017 Apr;64(2):343-358.
4. Curtis KM, Tepper NK, Jatlaoui TC, Berry-Bibee E, Horton LG, Zapata LB, Simmons KB, Pagano HP, Jamieson DJ, Whiteman MK. U.S. Medical Eligibility Criteria for Contraceptive Use, 2016. MMWR Recomm Rep. 2016 Jul 29;65(3):1-103.
5. Curtis KM, Jatlaoui TC, Tepper NK, Zapata LB, Horton LG, Jamieson DJ, Whiteman MK. U.S. Selected Practice Recommendations for Contraceptive Use, 2016. MMWR Recomm Rep. 2016 Jul 29;65(4):1-66.
6. World Health Organization. Medical Eligibility Criteria for Contraceptive Use 5th Ed. 2015. WHO Press (Switzerland).
7. Jaworek T, Xu H, Gaynor BJ, Cole JW, Rannikmae K, Stanne TM, Tomppo L, Abedi V, Amouyel P, Armstrong ND, Attia J, Bell S, Benavente OR, Boncoraglio GB, Butterworth A; Cervical Artery Dissections and Ischemic Stroke Patients (CADSIP) Consortium; Carcel-Marquez J, Chen Z, Chong

M, Cruchaga C, Cushman M, Danesh J, Debette S, Duggan DJ, Durda JP, Engstrom G, Enzinger C, Faul JD, Fecteau NS, Fernandez-Cadenas I, Gieger C, Giese AK, Grewal RP, Grittner U, Havulinna AS, Heitsch L, Hochberg MC, Holliday E, Hu J, Ilinca A; INVENT Consortium; Irvin MR, Jackson RD, Jacob MA, Rabionet R, Jimenez-Conde J, Johnson JA, Kamatani Y, Kardia SLR, Koido M, Kubo M, Lange L, Lee JM, Lemmens R, Levi CR, Li J, Li L, Lin K, Lopez H, Luke S, Maguire J, McArdle PF, McDonough CW, Meschia JF, Metso T, Müller-Nurasyid M, O'Connor TD, O'Donnell M, Peddareddygari LR, Pera J, Perry JA, Peters A, Putaala J, Ray D, Rexrode K, Ribases M, Rosand J, Rothwell PM, Rundek T, Ryan KA, Sacco RL, Salomaa V, Sanchez-Mora C, Schmidt R, Sharma P, Slowik A, Smith JA, Smith NL, Wassertheil-Smoller S, Söderholm M, Stine OC, Strbian D, Sudlow CLM, Tatlisumak T, Terao C, Thijs V, Torres-Aguila NP, Trégouët DA, Tuladhar AM, Veldink JH, Walters RG, Weir DR, Woo D, Worrall BB, Hong CC, Ross OA, Zand R, Leeuw FE, Lindgren AG, Pare G, Anderson CD, Markus HS, Jern C, Malik R, Dichgans M, Mitchell BD, Kittner SJ; Early Onset Stroke Genetics Consortium of the International Stroke Genetics Consortium (ISGC). Contribution of Common Genetic Variants to Risk of Early-Onset Ischemic Stroke. Neurology. 2022 Oct 17;99(16):e1738-e1754.

8. Asai Y, Baibergenova A, Dutil M, et al. Management of acne: Canadian clinical practice guideline. Appendix 5: Cost of drugs for each Canadian province and territory. CMAJ 2015. DOI:10.1503/cmaj.140665
9. Teen Health source. What is covered by OHIP+. Aug 2022. https://teenhealthsource.com/blog/what-is-covered-ohip-birth-control-edition/
10. Li F, Zhu L, Zhang J, He H, Qin Y, Cheng Y, Xie Z. Oral Contraceptive Use and Increased Risk of Stroke: A Dose-Response Meta-Analysis of Observational Studies. Front Neurol. 2019 Sep 23;10:993.

CHAPTER 10. INTRAUTERINE CONTRACEPTIVES

1. Black A, Guilbert E, Costescu D, Wagnet M, Whelan AM, et al. Canadian Contraception Consensus. SOGC Clinical Practice Guideline 2015: 37(10): 936-938.
2. United Nations, Department of Economic and Social Affairs, Population Division (2019). Contraceptive Use by Method 2019: Data Booklet.
3. Daniels K, Abma JC. Current contraceptive status among women aged 15–49: United States, 2017–2019. NCHS Data Brief, no 388. Hyattsville, MD: National Center for Health Statistics. 2020.
4. Rahman M, King C, Saikaly R, Sosa M, Sibaja K, Tran B, Tran S, Morello P, Yeon Seo S, Yeon Seo Y, Jacobs RJ. Differing Approaches to Pain Management for Intrauterine Device Insertion and Maintenance: A Scoping Review. Cureus. 2024 Mar 8;16(3):e55785.
5. Bahamondes L, Brache V, Meirik O, Ali M, Habib N, Landoulsi S; WHO Study Group on Contraceptive Implants for Women. A 3-year multicentre randomized controlled trial of etonogestrel- and levonorgestrel-releasing contraceptive implants, with non-randomized matched copper-intrauterine device controls. Hum Reprod. 2015 Nov;30(11):2527-38.
6. Sergison JE, Maldonado LY, Gao X, Hubacher D. Levonorgestrel intrauterine system associated amenorrhea: a systematic review and metaanalysis. Am J Obstet Gynecol. 2019 May;220(5):440-448.
7. Paré AK, Ouattara A, Yé D, Kabré B, Bako A, Abubakar BM, Kambou T. Management of Intrauterine Device Migrated into the Bladder: A Case Report and Literature Review. Case Rep Urol. 2020 Oct 31;2020:8850087.
8. Bürger Z, Bucher AM, Comasco E, Henes M, Hübner S, Kogler L, Derntl B. Association of levonorgestrel intrauterine devices with stress reactivity, mental health, quality of life and sexual functioning: A systematic review. Front Neuroendocrinol. 2021 Oct;63:100943.
9. Elsayed M, Dardeer KT, Khehra N, Padda I, Graf H, Soliman A, Makram AM, Zeiss R, Schönfeldt-Lecuona C. The potential association between psychiatric symptoms and the use of levonorgestrel intrauterine devices (LNG-IUDs): A systematic review. World J Biol Psychiatry. 2023 Jun-Jul;24(6):457-475.
10. Skovlund CW, Møller AL, Torp-Pedersen C, Mørch LS, Meaidi A. Depression risk in users of different doses of levonorgestrel intrauterine systems: a nationwide prospective cohort study. Lancet Reg Health Eur. 2024 Jan 5;38:100813.

11. Elsayed M, Dardeer KT, Khehra N, Padda I, Graf H, Soliman A, Makram AM, Zeiss R, Schönfeldt-Lecuona C. The potential association between psychiatric symptoms and the use of levonorgestrel intrauterine devices (LNG-IUDs): A systematic review. World J Biol Psychiatry. 2023 Jun-Jul;24(6):457-475.

CHAPTER 11. BODY LITERACY AND FERTILITY AWARENESS BASSED METHODS

1. Duane M, Stanford JB, Porucznik CA, Vigil P. Fertility Awareness-Based Methods for Women's Health and Family Planning. Front Med (Lausanne). 2022 May 24;9:858977
2. Yu JL, Su YF, Zhang C, Jin L, Lin XH, Chen LT, Huang HF, Wu YT. Tracking of menstrual cycles and prediction of the fertile window via measurements of basal body temperature and heart rate as well as machine-learning algorithms. Reprod Biol Endocrinol. 2022 Aug 13;20(1):118.
3. Black A, Guilbert E, Costescu D, Wagnet M, Whelan AM, et al. Canadian Contraception Consensus. SOGC Clinical Practice Guideline 2015: 37(10): 936-938.
4. Horvath S, Schreiber CA, Sonalkar S. Contraception. [Updated 2018 Jan 17]. In: Feingold KR, Anawalt B, Boyce A, et al., editors. Endotext [Internet]. South Dartmouth (MA): MDText.com, Inc.; 2000-.
5. Stanford JB, Duane M. Physicians Need Education About Fertility Awareness-Based Methods. Am Fam Physician. 2023 Sep;108(3):225-226.
6. Duane M, Martinez V, Berry M, Manhart MD. Evaluation of a fertility awareness-based shared decision-making tool part 1: Study design and impact on clinician knowledge. PEC Innov. 2022 Jun 28;1:100061.
7. Duane M, Martinez V, Berry M, Gilpatrick S, Manhart MD. Evaluation of a fertility awareness-based shared decision-making tool part 2: Patient experiences. PEC Innov. 2023 May 23;2:100169.
8. Davidson AV, Butler FM. Physician Dispositions Toward Noninvasive Non-Hormonal Contraception. Kans J Med. 2023 Apr 24;16:94-104.
9. Webb S, Cheng AL, Simmons R, Peragallo Urrutia R, Jennings V, Witt J. A Mixed-Methods Assessment of Health Care Providers' Knowledge, Attitudes, and Practices Around Fertility Awareness-Based Methods in Title X Clinics in the United States. Womens Health Rep (New Rochelle). 2020 Sep 15;1(1):354-365.

CHAPTER 12. OTHER CONTRACEPTIVES

1. Black A, Guilbert E, Costescu D, Wagnet M, Whelan AM, et al. Canadian Contraception Consensus. SOGC Clinical Practice Guideline 2015: 37(10): 936-938.
2. Moray KV, Chaurasia H, Sachin O, Joshi B. A systematic review on clinical effectiveness, side-effect profile and meta-analysis on continuation rate of etonogestrel contraceptive implant. Reprod Health. 2021 Jan 6;18(1):4.
3. Horvath S, Schreiber CA, Sonalkar S. Contraception. [Updated 2018 Jan 17]. In: Feingold KR, Anawalt B, Boyce A, et al., editors. Endotext [Internet]. South Dartmouth (MA): MDText.com, Inc.; 2000-.

CHAPTER 13. EMERGENCY CONTRACEPTION

1. Black A, Guilbert E, Costescu D, Wagnet M, Whelan AM, et al. Canadian Contraception Consensus. SOGC Clinical Practice Guideline 2015: 37(10): 936-938.
2. Edelman AB, Hennebold JD, Bond K, Lim JY, Cherala G, Archer DF, Jensen JT. Double Dosing Levonorgestrel-Based Emergency Contraception for Individuals With Obesity: A Randomized Controlled Trial. Obstet Gynecol. 2022 Jul 1;140(1):48-54.
3. Vogel L. Rethink weight limits on morning-after pill. CMAJ. 2015 Jul 14;187(10):719-720. doi: 10.1503/cmaj.109-5098.
4. Horvath S, Schreiber CA, Sonalkar S. Contraception. [Updated 2018 Jan 17]. In: Feingold KR, Anawalt B, Boyce A, et al., editors. Endotext [Internet]. South Dartmouth (MA): MDText.com, Inc.; 2000-.

CHAPTER 14. GENERAL MENSTRUAL HEALTH

1. Han L, Zhao T, Zhang R, Hao Y, Jiao M, Wu Q, Liu J, Zhou M. Burden of Nutritional Deficiencies in China: Findings from the Global Burden of Disease Study 2019. Nutrients. 2022 Sep 21;14(19):3919.
2. Canadian Association of Naturopathic Doctors, About Naturopathic Medicine, updated 2024. < https://www.cand.ca/about-naturopathic-medicine/>

CHAPTER 15. PAINFUL PERIODS

1. Stella Iacovides, Ingrid Avidon, Fiona C. Baker, What we know about primary dysmenorrhea today: a critical review, *Human Reproduction Update*. Nov/Dec 2015; 21(6): 762–778.
2. Gutke A, Sundfeldt K, De Baets L. Lifestyle and Chronic Pain in the Pelvis: State of the Art and Future Directions. J Clin Med. 2021 Nov 19;10(22):5397.
3. Hu Z, Tang L, Chen L, Kaminga AC, Xu H. Prevalence and risk factors associated with primary dysmenorrhea among Chinese female university students: a cross-sectional study. *J Pediatr Adolesc Gynecol*. 2020;33(1):15–22.
4. Pati GK, Kar C, Narayan J, Uthansingh K, Behera M, Sahu MK, Mishra D, Singh A. Irritable Bowel Syndrome and the Menstrual Cycle. Cureus. 2021 Jan 14;13(1):e12692.
5. Bajalan Z, Alimoradi Z, Moafi F. Nutrition as a Potential Factor of Primary Dysmenorrhea: A Systematic Review of Observational Studies. Gynecol Obstet Invest. 2019;84(3):209-224.
6. Helbig M., Vesper A.S., Beyer I., Fehm T. Does Nutrition Affect Endometriosis? Geburtshilfe Frauenheilkd. 2021;81:191–199. doi: 10.1055/a-1207-0557.
7. Zhu D, Xiao Y, Zhong G, Wei X, Wu J, Chen R, Jiao L. A Bibliometric Analysis of Acupuncture Therapy in the Treatment of Primary Dysmenorrhea from 2001 to 2021. J Pain Res. 2022 Sep 27;15:3043-3057.
8. Liu W, Wang CC, Lee KH, Ma X, Kang TL. Efficacy and Safety of Acupuncture and or Moxibustion for Managing Primary Dysmenorrhea: A Systematic Review and Meta-Analysis. Clin Nurs Res. 2022 Sep;31(7):1362-1375.
9. Ashrafi M, Jahangiri N, Jahanian Sadatmahalleh SH, Aliani F, Akhoond MR. Diet and The Risk of Endometriosis in Iranian Women: A Case-Control Study. Int J Fertil Steril. 2020 Oct;14(3):193-200.
10. Djuricic I, Calder PC. Beneficial Outcomes of Omega-6 and Omega-3 Polyunsaturated Fatty Acids on Human Health: An Update for 2021. Nutrients. 2021 Jul 15;13(7):2421.
11. Tsonis O, Gkrozou F, Barmpalia Z, Makopoulou A, Siafaka V. Integrating Lifestyle Focused Approaches into the Management of Primary Dysmenorrhea: Impact on Quality of Life. Int J Womens Health. 2021 Mar 17;13:327-336.
12. Li X, Hao X, Liu JH, Huang JP. Efficacy of non-pharmacological interventions for primary dysmenorrhoea: a systematic review and Bayesian network meta-analysis. BMJ Evid Based Med. 2024 Jan 19.
13. Mohammadi MM, Mirjalili R, Faraji A. The impact of omega-3 polyunsaturated fatty acids on primary dysmenorrhea: a systematic review and meta-analysis of randomized controlled trials. Eur J Clin Pharmacol 2022 May;78(5):721-731.
14. López-Liria R, Torres-Álamo L, Vega-Ramírez FA, García-Luengo AV, Aguilar-Parra JM, Trigueros-Ramos R, Rocamora-Pérez P. Efficacy of Physiotherapy Treatment in Primary Dysmenorrhea: A Systematic Review and Meta-Analysis. Int J Environ Res Public Health. 2021 Jul 23;18(15):7832.
15. Negi R, Sharma SK, Gaur R, Bahadur A, Jelly P. Efficacy of Ginger in the Treatment of Primary Dysmenorrhea: A Systematic Review and Meta-analysis. Cureus. 2021 Mar 6;13(3):e13743.
16. Rahnama P, Montazeri A, Huseini HF, Kianbakht S, Naseri M. Effect of Zingiber officinale R. rhizomes (ginger) on pain relief in primary dysmenorrhea: a placebo randomized trial. BMC Complement Altern Med. 2012 Jul 10;12:92.
17. Giti Ozgoli, Marjan Goli, and Fariborz Moattar. Comparison of Effects of Ginger, Mefenamic Acid, and Ibuprofen on Pain in Women with Primary Dysmenorrhea. The Journal of Alternative and Complementary Medicine 2009; 15:2, 129-132
18. Eshaghian R, Mazaheri M, Ghanadian M, Rouholamin S, Feizi A, Babaeian M. The effect of frankincense (Boswellia serrata, oleoresin) and ginger (Zingiber officinale, rhizoma) on heavy

menstrual bleeding: A randomized, placebo-controlled, clinical trial. Complement Ther Med. 2019 Feb;42:42-47.

19. Niazi A, Moradi M. The Effect of Chamomile on Pain and Menstrual Bleeding in Primary Dysmenorrhea: A Systematic Review. Int J Community Based Nurs Midwifery. 2021 Jul;9(3):174-186.
20. Samadipour E, Rakhshani MH, Kooshki A, Amin B. Local Usage of Nigella sativa Oil as an Innovative Method to Attenuate Primary Dysmenorrhea: A Randomized Double-blind Clinical Trial. Oman Med J. 2020 Sep 9;35(5):e167.
21. Sadeghi N, Paknezhad F, Rashidi Nooshabadi M, Kavianpour M, Jafari Rad S, Khadem Haghighian H. Vitamin E and fish oil, separately or in combination, on treatment of primary dysmenorrhea: a double-blind, randomized clinical trial. Gynecol Endocrinol. 2018 Sep;34(9):804-808.
22. Pakniat H, Chegini V, Ranjkesh F, Hosseini MA. Comparison of the effect of vitamin E, vitamin D and ginger on the severity of primary dysmenorrhea: a single-blind clinical trial. Obstet Gynecol Sci. 2019 Nov;62(6):462-468.
23. Md Amin NA, Sheikh Abdul Kadir SH, Arshad AH, Abdul Aziz N, Abdul Nasir NA, Ab Latip N. Are Vitamin E Supplementation Beneficial for Female Gynaecology Health and Diseases? Molecules. 2022 Mar 15;27(6):1896.

CHAPTER 16. PMS

1. Brzezinski AA, Wurtman JJ, Wurtman RJ, Gleason R, Greenfield J, Nader T. d-Fenfluramine suppresses the increased calorie and carbohydrate intakes and improves the mood of women with premenstrual depression. Obstet Gynecol. 1990 Aug;76(2):296-301.
2. Robertson, J. PMS & PMDD Webinar. The Confident Clinician Club. 2021 Feb.
3. Cerqueira RO, Frey BN, Leclerc E, Brietzke E. Vitex agnus castus for premenstrual syndrome and premenstrual dysphoric disorder: a systematic review. Arch Womens Ment Health. 2017 Dec;20(6):713-719.
4. Armour M, Ee CC, Hao J, Wilson TM, Yao SS, Smith CA. Acupuncture and acupressure for premenstrual syndrome. Cochrane Database Syst Rev. 2018 Aug 14;8(8):CD005290.
5. Mohammad Mehdi Mohammadi, Nahid Dehghan Nayeri, Monireh Mashhadi, Shokoh Varaei Effect of omega-3 fatty acids on premenstrual syndrome: A systematic review and meta-analysis. J Obstet Gynaecol Res. 2022 Jun;48(6):1293-1305.
6. Benton MJ, Hutchins AM, Dawes JJ. Effect of menstrual cycle on resting metabolism: A systematic review and meta-analysis. PLoS One. 2020 Jul 13;15(7):e0236025.
7. Weisz G, Knaapen L. Diagnosing and treating premenstrual syndrome in five western nations. Soc Sci Med. 2009 Apr;68(8):1498-505.
8. Takeda T, Tadakawa M, Koga S, Nagase S, Yaegashi N. Premenstrual symptoms and posttraumatic stress disorder in Japanese high school students 9 months after the great East-Japan earthquake. Tohoku J Exp Med. 2013 Jul;230(3):151-4.
9. Sims ST, Kerksick CM, Smith-Ryan AE, Janse de Jonge XAK, Hirsch KR, Arent SM, Hewlings SJ, Kleiner SM, Bustillo E, Tartar JL, Starratt VG, Kreider RB, Greenwalt C, Rentería LI, Ormsbee MJ, VanDusseldorp TA, Campbell BI, Kalman DS, Antonio J. International society of sports nutrition position stand: nutritional concerns of the female athlete. J Int Soc Sports Nutr. 2023 Dec;20(1):2204066.
10. Tartagni M, Cicinelli MV, Tartagni MV, Alrasheed H, Matteo M, Baldini D, De Salvia M, Loverro G, Montagnani M. Vitamin D Supplementation for Premenstrual Syndrome-Related Mood Disorders in Adolescents with Severe Hypovitaminosis D. J Pediatr Adolesc Gynecol. 2016 Aug;29(4):357-61.
11. Jarosz AC, El-Sohemy A. Association between Vitamin D Status and Premenstrual Symptoms. J Acad Nutr Diet. 2019 Jan;119(1):115-123.
12. Gök S, Gök B. Investigation of Laboratory and Clinical Features of Primary Dysmenorrhea: Comparison of Magnesium and Oral Contraceptives in Treatment. Cureus. 2022 Nov 29;14(11):e32028. doi: 10.7759/cureus.32028.
13. Bharati M. Comparing the Effects of Yoga & Oral Calcium Administration in Alleviating Symptoms of Premenstrual Syndrome in Medical Undergraduates. J Caring Sci. 2016 Sep 1;5(3):179-185.

14. Bahrami A, Avan A, Sadeghnia HR, Esmaeili H, Tayefi M, Ghasemi F, Nejati Salehkhani F, Arabpour-Dahoue M, Rastgar-Moghadam A, Ferns GA, Bahrami-Taghanaki H, Ghayour-Mobarhan M. High dose vitamin D supplementation can improve menstrual problems, dysmenorrhea, and premenstrual syndrome in adolescents. Gynecol Endocrinol. 2018 Aug;34(8):659-663.
15. Takeda T, Chiba Y. Evaluation of a natural S-equol supplement in treating premenstrual symptoms and the effect of the gut microbiota: An open-label pilot study. Neuropsychopharmacol Rep. 2022 Jun;42(2):127-134.
16. Witkoś J, Hartman-Petrycka M. The Influence of Running and Dancing on the Occurrence and Progression of Premenstrual Disorders. Int J Environ Res Public Health. 2021 Jul 27;18(15):7946.
17. Shobeiri F, Araste FE, Ebrahimi R, Jenabi E, Nazari M. Effect of calcium on premenstrual syndrome: A double-blind randomized clinical trial. Obstet Gynecol Sci. 2017 Jan;60(1):100-105.
18. Bryant M, Cassidy A, Hill C, Powell J, Talbot D, Dye L. Effect of consumption of soy isoflavones on behavioural, somatic and affective symptoms in women with premenstrual syndrome. Br J Nutr. 2005 May;93(5):731-9.
19. Thakur H, Pareek P, Sayyad MG, Otiv S. Association of Premenstrual Syndrome with Adiposity and Nutrient Intake Among Young Indian Women. Int J Womens Health. 2022 May 4;14:665-675.
20. Karimi Z, Dehkordi MA, Alipour A, Mohtashami T. Treatment of premenstrual syndrome: Appraising the effectiveness of cognitive behavioral therapy in addition to calcium supplement plus vitamin D. Psych J. 2018 Mar;7(1):41-50.
21. Arab A, Rafie N, Askari G, Taghiabadi M. Beneficial Role of Calcium in Premenstrual Syndrome: A Systematic Review of Current Literature. Int J Prev Med. 2020 Sep 22;11:156.
22. Thys-Jacobs S, McMahon D, Bilezikian JP. Cyclical changes in calcium metabolism across the menstrual cycle in women with premenstrual dysphoric disorder. J Clin Endocrinol Metab. 2007 Aug;92(8):2952-9. doi: 10.1210/jc.2006-2726.
23. Abdi F, Ozgoli G, Rahnemaie FA. A systematic review of the role of vitamin D and calcium in premenstrual syndrome. Obstet Gynecol Sci. 2019 Mar;62(2):73-86. doi: 10.5468/ogs.2019.62.2.73. Epub 2019 Feb 25. Erratum in: Obstet Gynecol Sci. 2020 Mar;63(2):213.
24. Whelan AM, Jurgens TM, Naylor H. Herbs, vitamins and minerals in the treatment of premenstrual syndrome: a systematic review. Can J Clin Pharmacol. 2009 Fall;16(3):e407-29.
25. Jang SH, Kim DI, Choi MS. Effects and treatment methods of acupuncture and herbal medicine for premenstrual syndrome/premenstrual dysphoric disorder: systematic review. BMC Complement Altern Med. 2014 Jan 10;14:11.
26. Khalesi ZB, Beiranvand SP, and Bokaie M. Efficacy of Chamomile in the Treatment of Premenstrual Syndrome: A Systematic Review. J Pharmacopuncture. 2019 Dec; 22(4): 204–209.
27. Quaranta S, Buscaglia MA, Meroni MG, Colombo E, Cella S. Pilot study of the efficacy and safety of a modified-release magnesium 250 mg tablet (Sincromag) for the treatment of premenstrual syndrome. Clin Drug Investig. 2007;27(1):51-8.
28. Boyle NB, Lawton C, Dye L. The Effects of Magnesium Supplementation on Subjective Anxiety and Stress-A Systematic Review. Nutrients. 2017 Apr 26;9(5):429. doi: 10.3390/nu9050429.
29. Retallick-Brown H, Blampied N, Rucklidge JJ. A Pilot Randomized Treatment-Controlled Trial Comparing Vitamin B6 with Broad-Spectrum Micronutrients for Premenstrual Syndrome. J Altern Complement Med. 2020 Feb;26(2):88-97.
30. Ebrahimi E, Khayati Motlagh S, Nemati S, Tavakoli Z. Effects of magnesium and vitamin b6 on the severity of premenstrual syndrome symptoms. J Caring Sci. 2012 Nov 22;1(4):183-9.
31. Zamani M, Neghab N, Torabian S. Therapeutic effect of Vitex agnus castus in patients with premenstrual syndrome. Acta Med Iran. 2012;50(2):101-6.
32. Schellenberg R, Zimmermann C, Drewe J, Hoexter G, Zahner C. Dose-dependent efficacy of the Vitex agnus castus extract Ze 440 in patients suffering from premenstrual syndrome. Phytomedicine. 2012 Nov 15;19(14):1325-31.

CHAPTER 17. PMDD

1. Eisenlohr-Moul TA, Girdler SS, Schmalenberger KM, Dawson DN, Surana P, Johnson JL, Rubinow DR (January 2017). "Toward the Reliable Diagnosis of DSM-5 Premenstrual Dysphoric Disorder:

The Carolina Premenstrual Assessment Scoring System (C-PASS)". *The American Journal of Psychiatry*. **174** (1): 51–59.
2. Reid RL. Premenstrual Dysphoric Disorder (Formerly Premenstrual Syndrome) [Updated 2017 Jan 23]. In: Feingold KR, Anawalt B, Boyce A, et al., editors. Endotext [Internet]. South Dartmouth (MA): MDText.com, Inc.; 2000-. Table 1, Diagnostic Criteria for Premenstrual Dysphoric Disorder (PMDD).
3. Pearlstein T. Bipolar Disorder and Premenstrual Dysphoric Disorder: Comorbidity Conundrum. J Clin Psychiatry. 2022 Oct 26;83(6):22com14573.
4. Lin PC, Ko CH, Yen JY. Early and Late Luteal Executive Function, Cognitive and Somatic Symptoms, and Emotional Regulation of Women with Premenstrual Dysphoric Disorder. J Pers Med. 2022 May 18;12(5):819. doi: 10.3390/jpm12050819.
5. Mishra S, Elliott H, Marwaha R. Premenstrual Dysphoric Disorder. [Updated 2022 May 5]. In: StatPearls [Internet]. Treasure Island (FL): StatPearls Publishing; 2022 Jan-.
6. Dubol M, Stiernman L, Wikström J, Lanzenberger R, Neill Epperson C, Sundström-Poromaa I, Bixo M, Comasco E. Differential grey matter structure in women with premenstrual dysphoric disorder: evidence from brain morphometry and data-driven classification. Transl Psychiatry. 2022 Jun 15;12(1):250.
7. Cunningham J, Yonkers KA, O'Brien S, Eriksson E. Update on research and treatment of premenstrual dysphoric disorder. Harv Rev Psychiatry. 2009;17(2):120-37.
8. Takeda T, Tadakawa M, Koga S, Nagase S, Yaegashi N. Premenstrual symptoms and posttraumatic stress disorder in Japanese high school students 9 months after the great East-Japan earthquake. Tohoku J Exp Med. 2013 Jul;230(3):151-4.
9. Frey BN, Allega OR, Eltayebani M, Syan SK, Mendes-Ribeiro J, Minuzzi L. A DSM-5-based tool to monitor concurrent mood and premenstrual symptoms: the McMaster Premenstrual and Mood Symptom Scale (MAC-PMSS). BMC Womens Health. 2022 Mar 30;22(1):96.
10. Morishita C, Inoue T, Honyashiki M, Ono M, Iwata Y, Tanabe H, Kusumi I, Masuya J. Roles of childhood maltreatment, personality traits, and life stress in the prediction of severe premenstrual symptoms. Biopsychosoc Med. 2022 Apr 28;16(1):11.
11. Naguy A, El-Sheshai A, Thiguti SH, Alamiri B. Psychopharmacotherapy of Premenstrual Dysphoric Disorder-*New Vistas*. Psychopharmacol Bull. 2022 Jun 27;52(3):81-83.
12. Sundblad C, Wikander I, Andersch B, Eriksson E. A naturalistic study of paroxetine in premenstrual syndrome: efficacy and side-effects during ten cycles of treatment. Eur Neuropsychopharmacol. 1997 Aug;7(3):201-6.
13. Brzezinski AA, Wurtman JJ, Wurtman RJ, Gleason R, Greenfield J, Nader T. d-Fenfluramine suppresses the increased calorie and carbohydrate intakes and improves the mood of women with premenstrual depression. Obstet Gynecol. 1990 Aug;76(2):296-301.
14. Cerqueira RO, Frey BN, Leclerc E, Brietzke E. Vitex agnus castus for premenstrual syndrome and premenstrual dysphoric disorder: a systematic review. Arch Womens Ment Health. 2017 Dec;20(6):713-719.
15. Jang SH, Kim DI, Choi MS. Effects and treatment methods of acupuncture and herbal medicine for premenstrual syndrome/premenstrual dysphoric disorder: systematic review. BMC Complement Altern Med. 2014 Jan 10;14:11.
16. Rajabi F, Rahimi M, Sharbafchizadeh MR, Tarrahi MJ. Saffron for the Management of Premenstrual Dysphoric Disorder: A Randomized Controlled Trial. Adv Biomed Res. 2020 Oct 30;9:60.
17. Carvalho F, Weires K, Ebling M, Padilha Mde S, Ferrão YA, Vercelino R. Effects of acupuncture on the symptoms of anxiety and depression caused by premenstrual dysphoric disorder. Acupunct Med. 2013 Dec;31(4):358-63.
18. Taguchi R, Matsubara S, Yoshimoto S, Imai K, Ohkuchi A, Kitakoji H. Acupuncture for premenstrual dysphoric disorder. Arch Gynecol Obstet. 2009 Dec;280(6):877-81.

CHAPTER 18. ABNORMAL UTERINE BLEEDING

1. Munro MG, Critchley HOD, Fraser IS., FIGO Menstrual Disorders Committee. The two FIGO systems for normal and abnormal uterine bleeding symptoms and classification of causes of abnormal

uterine bleeding in the reproductive years: 2018 revisions. Int J Gynaecol Obstet. 2018 Dec;143(3):393-408.

2. Jain V, Chodankar RR, Maybin JA, Critchley HOD. Uterine bleeding: how understanding endometrial physiology underpins menstrual health. Nat Rev Endocrinol. 2022 May;18(5):290-308.
3. Chodankar R and Critchley HOD. Biomarkers in abnormal uterine bleeding. Biol Reprod. 2019 Dec; 101(6): 1155–1166.
4. Davis E, Sparzak PB. Abnormal Uterine Bleeding. [Updated 2022 Sep 9]. In: StatPearls [Internet]. Treasure Island (FL): StatPearls Publishing; 2022 Jan-.
5. Selter JH, Price TM, Harris BS. Fibroids in pregnancy: a growing or shrinking issue? Fertil Steril. 2022 Oct;118(4):666-667.
6. W Meng, L Lao, ZJ Zhang, WL Lin, YB Zhang, WF Yeung, YMB Yu, HYE Ng, JP Chen, J Su, JH Rong, PYF Lam, E Lee. Tumour-shrinking decoction for symptomatic uterine fibroids: a double-blind, randomised, two-dose trial (abridged secondary publication). Hong Kong Med J 2021;27(Suppl 2):S8-10.
7. Al-Husban N, Odeh O, AlRamahi M, Qadri S, Al-Husban H. Fertility-enhancing hysteroscopic surgery; multi-center retrospective cohort study of reproductive outcome. BMC Womens Health. 2023 Aug 29;23(1):459.
8. Jadhav SB, Shah N, Rathi A, Rathi V, Rathi A. Serratiopeptidase: Insights into the therapeutic applications. Biotechnol Rep (Amst). 2020 Oct 17;28:e00544.

CHAPTER 19. PCOS

1. Teede HJ, Tay CT, Laven JJE, Dokras A, Moran LJ, Piltonen TT, Costello MF, Boivin J, Redman LM, Boyle JA, Norman RJ, Mousa A, Joham AE; International PCOS Network. Recommendations From the 2023 International Evidence-based Guideline for the Assessment and Management of Polycystic Ovary Syndrome. J Clin Endocrinol Metab. 2023 Aug 15
2. Shahbaz M, Almatooq H, Foucambert P, Esbrand FD, Zafar S, Panthangi V, Cyril Kurupp AR, Raju A, Luthra G, Khan S. A Systematic Review of the Risk of Non-alcoholic Fatty Liver Disease in Women With Polycystic Ovary Syndrome. Cureus. 2022 Oct 4;14(10):e29928.
3. Deswal R, Narwal V, Dang A, Pundir CS. The Prevalence of Polycystic Ovary Syndrome: A Brief Systematic Review. J Hum Reprod Sci. 2020 Oct-Dec;13(4):261-271. doi: 10.4103/jhrs.JHRS_95_18. Epub 2020 Dec 28.
4. Lujan ME, Chizen DR, Pierson RA. Diagnostic criteria for polycystic ovary syndrome: pitfalls and controversies. J Obstet Gynaecol Can. 2008 Aug;30(8):671-679.
5. Armanini D, Boscaro M, Bordin L, Sabbadin C. Controversies in the Pathogenesis, Diagnosis and Treatment of PCOS: Focus on Insulin Resistance, Inflammation, and Hyperandrogenism. Int J Mol Sci. 2022 Apr 8;23(8):4110.
6. Palomba S, Piltonen TT, Giudice LC. Endometrial function in women with polycystic ovary syndrome: a comprehensive review. Hum Reprod Update. 2021 Apr 21;27(3):584-618.
7. Lee IT, Sansone S, Irfan M, Copp T, Beidas R, Dokras A. Implementation of international guidelines for polycystic ovary syndrome: barriers and facilitators among gynecologists and primary care providers. F S Rep. 2022 Jan 25;3(2):94-101. doi: 10.1016/j.xfre.2022.01.005.
8. Hosseinzadeh P, Barsky M, Gibbons WE, Blesson CS. Polycystic Ovary Syndrome and the Forgotten Uterus. F S Rev. 2021 Jan;2(1):11-20.
9. Dinicola S, Unfer V, Facchinetti F, Soulage CO, Greene ND, Bizzarri M, Laganà AS, Chan SY, Bevilacqua A, Pkhaladze L, Benvenga S, Stringaro A, Barbaro D, Appetecchia M, Aragona C, Bezerra Espinola MS, Cantelmi T, Cavalli P, Chiu TT, Copp AJ, D'Anna R, Dewailly D, Di Lorenzo C, Diamanti-Kandarakis E, Hernández Marín I, Hod M, Kamenov Z, Kandaraki E, Monastra G, Montanino Oliva M, Nestler JE, Nordio M, Ozay AC, Papalou O, Porcaro G, Prapas N, Roseff S, Vazquez-Levin M, Vucenik I, Wdowiak A. Inositols: From Established Knowledge to Novel Approaches. Int J Mol Sci. 2021 Sep 30;22(19):10575.
10. Oguz SH, Yildiz BO. An Update on Contraception in Polycystic Ovary Syndrome. Endocrinol Metab (Seoul). 2021 Apr;36(2):296-311.

11. Hanem LGE, Salvesen Ø, Juliusson PB, et al. Intrauterine metformin exposure and offspring cardiometabolic risk factors (PedMet study): a 5-10 year follow-up of the PregMet randomised controlled trial. Lancet Child Adolesc Health 2019;3:166-74. 10.1016/S2352-4642(18)30385-7
12. Zhu D, Chen Y, Huang J, Deng H, Shen X, Lu D, Xu L. Effects of metformin on pregnancy outcome, metabolic profile, and sex hormone levels in women with polycystic ovary syndrome and their offspring: a systematic review and meta-analysis. Ann Transl Med. 2022 Apr;10(7):418.
13. Mercorio A, Della Corte L, De Angelis MC, Buonfantino C, Ronsini C, Bifulco G, Giampaolino P. Ovarian Drilling: Back to the Future. Medicina (Kaunas). 2022 Jul 27;58(8):1002.
14. Flore G, Preti A, Carta MG, Deledda A, Fosci M, Nardi AE, Loviselli A, Velluzzi F. Weight Maintenance after Dietary Weight Loss: Systematic Review and Meta-Analysis on the Effectiveness of Behavioural Intensive Intervention. Nutrients. 2022 Mar 16;14(6):1259.
15. Greenway FL. Physiological adaptations to weight loss and factors favouring weight regain. Int J Obes (Lond). 2015 Aug;39(8):1188-96.
16. Gordon, Aubrey. "What We Don't Talk About When We Talk About Fat." Beacon Press: 2020.
17. Pal L, Zhang H, Williams J, Santoro NF, Diamond MP, Schlaff WD, Coutifaris C, Carson SA, Steinkampf MP, Carr BR, McGovern PG, Cataldo NA, Gosman GG, Nestler JE, Myers E, Legro RS; Reproductive Medicine Network. Vitamin D Status Relates to Reproductive Outcome in Women With Polycystic Ovary Syndrome: Secondary Analysis of a Multicenter Randomized Controlled Trial. J Clin Endocrinol Metab. 2016 Aug;101(8):3027-35.
18. Zarezadeh M, Dehghani A, Faghfouri AH, Radkhah N, Naemi Kermanshahi M, Hamedi Kalajahi F, Mohammadzadeh Honarvar N, Ghoreishi Z, Ostadrahimi A, Ebrahimi Mamaghani M. Inositol supplementation and body mass index: A systematic review and meta-analysis of randomized clinical trials. Obes Sci Pract. 2021 Oct 22;8(3):387-397.
19. Soldat-Stankovic V, Popovic Pejicic S, Stankovic S, Jovanic J, Bjekic-Macut J, Livadas S, Ognjanovic S, Mastorakos G, Micic D, Macut D. The effect of myoinositol and metformin on cardiovascular risk factors in women with polycystic ovary syndrome: a randomized controlled trial. Acta Endocrinol (Buchar). 2021 Apr-Jun;17(2):241-247.
20. Zhao H, Xing C, Zhang J, He B. Comparative efficacy of oral insulin sensitizers metformin, thiazolidinediones, inositol, and berberine in improving endocrine and metabolic profiles in women with PCOS: a network meta-analysis. Reprod Health. 2021 Aug 18;18(1):171.
21. Patten RK, McIlvenna LC, Levinger I, Garnham AP, Shorakae S, Parker AG, McAinch AJ, Rodgers RJ, Hiam D, Moreno-Asso A, Stepto NK. High-intensity training elicits greater improvements in cardio-metabolic and reproductive outcomes than moderate-intensity training in women with polycystic ovary syndrome: a randomized clinical trial. Hum Reprod. 2022 May 3;37(5):1018-1029.
22. Mishra N, Verma R, Jadaun P. Study on the Effect of Berberine, Myoinositol, and Metformin in Women with Polycystic Ovary Syndrome: A Prospective Randomised Study. Cureus. 2022 Jan 31;14(1):e21781.
23. Sandhu JK, Waqar A, Jain A, Joseph C, Srivastava K, Ochuba O, Alkayyali T, Ruo SW, Poudel S. Oxidative Stress in Polycystic Ovarian Syndrome and the Effect of Antioxidant N-Acetylcysteine on Ovulation and Pregnancy Rate. Cureus. 2021 Sep 11;13(9):e17887.
24. Thakker D, Raval A, Patel I, Walia R. N-acetylcysteine for polycystic ovary syndrome: a systematic review and meta-analysis of randomized controlled clinical trials. Obstet Gynecol Int. 2015;2015:817849.
25. Wei W, Zhao H, Wang A, Sui M, Liang K, Deng H, Ma Y, Zhang Y, Zhang H, Guan Y. A clinical study on the short-term effect of berberine in comparison to metformin on the metabolic characteristics of women with polycystic ovary syndrome. Eur J Endocrinol. 2012 Jan;166(1):99-105.
26. Armanini D, Castello R, Scaroni C, Bonanni G, Faccini G, Pellati D, Bertoldo A, Fiore C, Moghetti P. Treatment of polycystic ovary syndrome with spironolactone plus licorice. Eur J Obstet Gynecol Reprod Biol. 2007 Mar;131(1):61-67.
27. Arentz S, Abbott JA, Smith CA, Bensoussan A. Herbal medicine for the management of polycystic ovary syndrome (PCOS) and associated oligo/amenorrhoea and hyperandrogenism; a review of the laboratory evidence for effects with corroborative clinical findings. BMC Complement Altern Med. 2014 Dec 18;14:511.

28. Mavranezouli I, Daly CH, Welton NJ, Deshpande S, Berg L, Bromham N, Arnold S, Phillippo DM, Wilcock J, Xu J, Ravenscroft JC, Wood D, Rafiq M, Fou L, Dworzynski K, Healy E. A systematic review and network meta-analysis of topical pharmacological, oral pharmacological, physical and combined treatments for acne vulgaris. Br J Dermatol. 2022 Nov;187(5):639-649.
29. Glass GE. Photobiomodulation: The Clinical Applications of Low-Level Light Therapy. Aesthet Surg J. 2021 May 18;41(6):723-738.
30. Gupta AK, Quinlan EM, Venkataraman M, Bamimore MA. Microneedling for Hair Loss. J Cosmet Dermatol. 2022 Jan;21(1):108-117.
31. Zhou Y, Chen C, Qu Q, Zhang C, Wang J, Fan Z, Miao Y, Hu Z. The effectiveness of combination therapies for androgenetic alopecia: A systematic review and meta-analysis. Dermatol Ther. 2020 Jul;33(4):e13741.

CHAPTER 20. ENDOMETRIOSIS

1. Ottolina J, Schimberni M, Makieva S, Bartiromo L, Fazia T, Bernardinelli L, Viganò P, Candiani M, Gentilini D. Early-life factors, in-utero exposures and endometriosis risk: a meta-analysis. Reprod Biomed Online. 2020 Aug;41(2):279-289.
2. Taylor HS, Kotlyar AM, Flores VA. Endometriosis is a chronic systemic disease: clinical challenges and novel innovations. Lancet. 2021 Feb 27;397(10276):839-852.
3. Pirtea P, Vulliemoz N, de Ziegler D, Ayoubi JM. Infertility workup: identifying endometriosis. Fertil Steril. 2022 Jul;118(1):29-33.
4. Wu XG, Chen JJ, Zhou HL, Wu Y, Lin F, Shi J, Wu HZ, Xiao HQ, Wang W. Identification and Validation of the Signatures of Infiltrating Immune Cells in the Eutopic Endometrium Endometria of Women With Endometriosis. Front Immunol. 2021 Sep 3;12:671201.
5. Habib N, Buzzaccarini G, Centini G, Moawad GN, Ceccaldi PF, Gitas G, Alkatout I, Gullo G, Terzic S, Sleiman Z. Impact of lifestyle and diet on endometriosis: a fresh look to a busy corner. Prz Menopauzalny. 2022 Jun;21(2):124-132.
6. Redwine DB. Was Sampson wrong? Fertil Steril. 2002 Oct;78(4):686-93.
7. Mak J, Leonardi M, Condous G. 'Seeing is believing': arguing for diagnostic laparoscopy as a diagnostic test for endometriosis. Reprod Fertil. 2022 Jun 10;3(3):C23-C28.
8. MRI in the Diagnosis of Endometriosis and Related Diseases. Korean J Radiol. 2022 Apr;23(4):426-445. doi: 10.3348/kjr.2021.0405.
9. Wang L, Sun J. ASPN Is a Potential Biomarker and Associated with Immune Infiltration in Endometriosis. Genes (Basel). 2022 Jul 28;13(8):1352.
10. Micu R, Gaia-Oltean AMI, Budişan L, Braicu C, Irimie A, Berindan-Neagoe I. The added value of CA125, HE4, and CA72-4 as markers for ovarian endometriosis diagnosis. Rom J Morphol Embryol. 2023 Apr-Jun;64(2):159-164.
11. Uzuner C, Mak J, El-Assaad F, Condous G. The bidirectional relationship between endometriosis and microbiome. Front Endocrinol (Lausanne). 2023 Mar 7;14:1110824.
12. Nabi MY, Nauhria S, Reel M, Londono S, Vasireddi A, Elmiry M, Ramdass PVAK. Endometriosis and irritable bowel syndrome: A systematic review and meta-analyses. Front Med (Lausanne). 2022 Jul 25;9:914356. doi: 10.3389/fmed.2022.914356.
13. *Below the Belt*. Cohn, Shannon. 2023. <https://www.belowthebelt.film/>
14. Hung SW, Zhang R, Tan Z, Chung JPW, Zhang T, Wang CC. Pharmaceuticals targeting signaling pathways of endometriosis as potential new medical treatment: A review. Med Res Rev. 2021 Jul;41(4):2489-2564.
15. Ashrafi M, Jahangiri N, Jahanian Sadatmahalleh SH, Aliani F, Akhoond MR. Diet and The Risk of Endometriosis in Iranian Women: A Case-Control Study. Int J Fertil Steril. 2020 Oct;14(3):193-200.
16. Matta K, Koual M, Ploteau S, Coumoul X, Audouze K, Le Bizec B, Antignac JP, Cano-Sancho G. Associations between Exposure to Organochlorine Chemicals and Endometriosis: A Systematic Review of Experimental Studies and Integration of Epidemiological Evidence. Environ Health Perspect. 2021 Jul;129(7):76003.
17. Gutke A, Sundfeldt K, De Baets L. Lifestyle and Chronic Pain in the Pelvis: State of the Art and Future Directions. J Clin Med. 2021 Nov 19;10(22):5397.

18. Jurkiewicz-Przondziono J, Lemm M, Kwiatkowska-Pamuła A, Ziółko E, Wójtowicz MK. Influence of diet on the risk of developing endometriosis. Ginekol Pol. 2017;88(2):96-102.
19. Helbig M., Vesper A.S., Beyer I., Fehm T. Does Nutrition Affect Endometriosis? Geburtshilfe Frauenheilkd. 2021;81:191–199. doi: 10.1055/a-1207-0557.
20. Chiarle G, Allais G, Sinigaglia S, Airola G, Rolando S, Bergandi F, Micalef S, Benedetto C. Acupuncture for pain and pain-related disability in deep infiltrating endometriosis. Front Pain Res (Lausanne). 2024 Mar 8;5:1279312.
21. Kong S, Zhang YH, Liu CF, Tsui I, Guo Y, Ai BB, Han FJ. The complementary and alternative medicine for endometriosis: a review of utilization and mechanism. Evid Based Complement Alternat Med. 2014;2014:146383.
22. Santanam N, Kavtaradze N, Murphy A, Dominguez C, Parthasarathy S. Antioxidant supplementation reduces endometriosis-related pelvic pain in humans. Transl Res. 2013 Mar;161(3):189-95.
23. Indraccolo U, Favilli A, Dell'Anna A, Di Francesco A, Dionisi B, Giugliano E, Murina F, Stocco E. Looking for Responders among Women with Chronic Pelvic Pain Treated with a Comicronized Formulation of Micronized Palmitoylethanolamide and Polydatin. Biomed Res Int. 2022 May 7;2022:8620077.
24. Gołąbek A, Kowalska K, Olejnik A. Polyphenols as a Diet Therapy Concept for Endometriosis-Current Opinion and Future Perspectives. Nutrients. 2021 Apr 18;13(4):1347.
25. Salmeri N, Sinagra E, Dolci C, Buzzaccarini G, Sozzi G, Sutera M, Candiani M, Ungaro F, Massimino L, Danese S, Mandarino FV. Microbiota in Irritable Bowel Syndrome and Endometriosis: Birds of a Feather Flock Together-A Review. Microorganisms. 2023 Aug 15;11(8):2089.

CHAPTER 21. OTHER MENSTRUAL AND HORMONAL CONCERNS

1. Fontana L, Garzia E, Marfia G, Galiano V, Miozzo M. Epigenetics of functional hypothalamic amenorrhea. Front Endocrinol (Lausanne). 2022 Aug 11;13:953431.Mobeen S, Apostol R. Ovarian Cyst. [Updated 2022 Jun 13]. In: StatPearls [Internet]. Treasure Island (FL): StatPearls Publishing; 2022 Jan-.
2. Guile SL, Mathai JK. Ovarian Torsion. [Updated 2022 Jul 18]. In: StatPearls [Internet]. Treasure Island (FL): StatPearls Publishing; 2022 Jan-.
3. Kota AS, Ejaz S. Precocious Puberty. [Updated 2022 Jul 4]. In: StatPearls [Internet]. Treasure Island (FL): StatPearls Publishing; 2022 Jan-.
4. Aguirre RS, Eugster EA. Central precocious puberty: From genetics to treatment. Best Pract Res Clin Endocrinol Metab. 2018 Aug;32(4):343-354.
5. Verrilli L, Johnstone E, Welt C, Allen-Brady K. Primary ovarian insufficiency has strong familiality: results of a multigenerational genealogical study. Fertil Steril. 2022 Oct 22:S0015-0282(22)01444-3.
6. Cai WY, Luo X, Wu W, Song J, Xie NN, Duan C, Wu XK, Xu J. Metabolic differences in women with premature ovarian insufficiency: a systematic review and meta-analysis. J Ovarian Res. 2022 Sep 30;15(1):109.
7. The ESHRE Guideline Group on POI, L. Webber, M. Davies, R. Anderson, J. Bartlett, D. Braat, B. Cartwright, R. Cifkova, S. de Muinck Keizer-Schrama, E. Hogervorst, F. Janse, L. Liao, V. Vlaisavljevic, C. Zillikens, N. Vermeulen, ESHRE Guideline: management of women with premature ovarian insufficiency, *Human Reproduction*, Volume 31, Issue 5, May 2016, Pages 926–937

CHAPTER 22: PERIODS AND POLITICS

1. Crotti R, Pal Kusum K, Ratcheva V, Zahidi S. Global Gender Gap Report. World Economic Forum. 2021. https://www.weforum.org/reports/global-gender-gap-report-2021
2. Bryant T, Raphael D, Shrecker T, Labonte R. Canada: A land of missed opportunity for addressing the social determinants of health. Health Policy 2011; 101(1): 44-58.
3. Sultana A. Patriarchy and Women's Subordination: A Theoretical Analysis. The Arts Faculty Journal, University of Dhaka. July 2010-June 2011.
4. Williams DR. Stress and the Mental Health of Populations of Color: Advancing Our Understanding of Race-related Stressors. J Health Soc Behav. 2018 Dec;59(4):466-485.

5. Johnston J, Cushing L. Chemical Exposures, Health, and Environmental Justice in Communities Living on the Fenceline of Industry. Curr Environ Health Rep. 2020 Mar;7(1):48-57.
6. O'Flynn N. Menstrual symptoms: the importance of social factors in women's experiences. Br J Gen Pract. 2006 Dec;56(533):950-7. Erratum in: Br J Gen Pract. 2007 Feb;57(535):156.
7. Gaybor J, Harcourt W. Seeing the colour red: menstruation in global body politics. Global Public Health. 2022; 17(10): 2388-2400.
8. Rossouw L, Ross H. Understanding Period Poverty: Socio-Economic Inequalities in Menstrual Hygiene Management in Eight Low- and Middle-Income Countries. Int J Environ Res Public Health. 2021 Mar 4;18(5):2571.

CHAPTER 23. MENSTRUATION AND MEDICAL INDUSTRY

1. National Institutes of Health. NIH Inclusion Outreach toolkit: How to Engage, Recruit, and Retain Women in Clinical Research. Updated 2022. <https://orwh.od.nih.gov/toolkit/recruitment/history>
2. Talumaa B, Brown A, Batterham RL, Kalea AZ. Effective strategies in ending weight stigma in healthcare. Obes Rev. 2022 Oct;23(10):e13494.
3. Cénat JM, Kouamou LN, Moshirian Farahi SMM, Darius WP, Dalexis RD, Charles M, Kogan CS. Perceived racial discrimination, psychosomatic symptoms, and resilience among Black individuals in Canada: A moderated mediation model. J Psychosom Res. 2022 Oct 7;163:111053.
4. Barrientos HFK. The Intersectionality of Race, Innovative Medical Research, and Eugenics. University of Colorado Boulder. June 7, 2022. <https://www.colorado.edu/herbst/2022/06/07/intersectionality-race-innovative-medical-research-and-eugenics>
5. Mahabir, D.F., O'Campo, P., Lofters, A. *et al.* Experiences of everyday racism in Toronto's health care system: a concept mapping study. *Int J Equity Health* **20**, 74 (2021).
6. World Health Organization Constitution of the World Health Organization. 1946. <https://www.who.int/about/accountability/governance/constitution>
7. National Wellness Institute. NWI's Six Dimensions of Wellness. 2023. <https://nationalwellness.org/resources/six-dimensions-of-wellness/>
8. Global Wellness Institute. What is Wellness? FL, USA. 2023. <https://globalwellnessinstitute.org/what-is-wellness/>
9. Ridd M, Shaw A, Lewis G, Salisbury C. The patient-doctor relationship: a synthesis of the qualitative literature on patients' perspectives. Br J Gen Pract. 2009 Apr;59(561):e116-33.
10. Daundasekara SS, Arlinghaus KR, Johnston CA. The Importance of Language in Behavior Change. Am J Lifestyle Med. 2019 Feb 19;13(3):239-242.
11. Dionisio K. Instagram Post: @drkarand, Mar 13, 2023. https://www.instagram.com/p/Cpvz9gUtx5S/?igshid=MzRlODBiNWFlZA==
12. Zelin NS, Scott C, Avila-Quintero VJ, Curlin K, Flores JM, Bloch MH. Sexual Orientation and Racial Bias in Relation to Medical Specialty. J Homosex. 2022 Oct 21:1-26.
13. Soled KRS, Dimant OE, Tanguay J, Mukerjee R, Poteat T. Interdisciplinary clinicians' attitudes, challenges, and success strategies in providing care to transgender people: a qualitative descriptive study. BMC Health Serv Res. 2022 Sep 8;22(1):1134.
14. Schatz AA, Brooks-Coley K, Harrington E, Murray MS, Carlson RW. Patient, Caregiver, and Oncologist Experiences With and Perceptions of Racial Bias and Discrimination in Cancer Care Delivery. J Natl Compr Canc Netw. 2022 Oct;20(10):1092-1098.
15. De Vries ALC, Cohen-Kettenis PT. Clinical management of gender dysphoria in children and adolescents: the Dutch approach. J Homosex. 2012; 59: 301.
16. Schwartz BI, Effron A, Bear B, Eisenberg J, Felleman S, Kazak AE. Experiences with menses in transgender and gender nonbinary adolescents. J Ped Adolesc Gyn.
17. Mittone DF, Bailey CP, Eddy EL, Napolitano MA, Vyas A. Women's Satisfaction With Telehealth Services During The COVID-19 Pandemic: Cross-sectional Survey Study. JMIR Pediatr Parent. 2022 Oct 14;5(4):e41356.

18. O'Flynn N. Menstrual symptoms: the importance of social factors in women's experiences. Br J Gen Pract. 2006 Dec;56(533):950-7. Erratum in: Br J Gen Pract. 2007 Feb;57(535):156.

CHAPTER 24. THE P WORD

1. Brown B. Daring Greatly: How the Courage to Be Vulnerable Transforms the Way We Live, Love, Parent, and Lead. 2013. London, England: Portfolio Penguin.
2. Gharib M. Why 2015 Was the Year of the Period, and we Don't Mean Punctuation. Dec 31, 2015. NPR. <https://www.npr.org/sections/health-shots/2015/12/31/460726461/why-2015-was-the-year-of-the-period-and-we-dont-mean-punctuation>
3. Jurić Petričević S, Buljan I, Bjelanović D, Mrduljaš-Đujić N, Pekez T, Ćurković M, Vojvodić Ž, Pavličević I, Marušić M, Marušić A. Effectiveness of letters to patients with or without Cochrane blogshots on 10-year cardiovascular risk change among women in menopausal transition: 6-month three-arm randomized controlled trial. BMC Med. 2022 Oct 20;20(1):381.
4. Gottlieb A. Menstrual Taboos: Moving Beyond the Curse. 2020 Jul 25. In: Bobel C, Winkler IT, Fahs B, et al., editors. The Palgrave Handbook of Critical Menstruation Studies [Internet]. Singapore: Palgrave Macmillan; 2020. Chapter 14. Available from: https://www.ncbi.nlm.nih.gov/books/NBK565616/ doi: 10.1007/978-981-15-0614-7_14
5. Buckley T, Gottlieb A. A critical appraisal of theories of menstrual symbolism. In: Buckley T, Gottlieb A, editors. Blood magic. The anthropology of menstruation. Berkeley: University of California Press; 1988. pp. 3–50.
6. Houppert K, The Curse: Confronting the Last Unmentionable Taboo: Menstruation. Farrar, Straus and Giroux, 1999.
7. Mooney-Somers J, Perz J, Ussher JM. A complex negotiation: women's experiences of naming and not naming premenstrual distress in couple relationships. Women Health. 2008;47(3):57-77. doi: 10.1080/03630240802134134. Erratum in: Women Health. 2009 Mar-May;49(2-3):262. PMID: 18714712.
8. O'Flynn N. Menstrual symptoms: the importance of social factors in women's experiences. Br J Gen Pract. 2006 Dec;56(533):950-7. Erratum in: Br J Gen Pract. 2007 Feb;57(535):156.
9. Oinas E. Medicalisation by whom? Accounts of menstruation conveyed by young women and medical experts in advisory columns. Sociol Health Illn. 1998;20:52–70.
10. Kissling EA. Bleeding out loud: communication about menstruation. Fem Psychol. 1996;6(4):481–504.

CHAPTER 25. PERIOD PRODUCTS

1. Canada Labour Code. Regulations Amending Certain Regulations Made Under the Canada Labour Code (Menstrual Products). Canada Gazette. 2022; Part I (156): Number 42. <https://canadagazette.gc.ca/rp-pr/p1/2022/2022-10-15/html/reg1-eng.html>
2. Liberal Party of Canada. Free Menstrual Products. 2022. <https://liberal.ca/our-platform/free-menstrual-products/>
3. DeVries AS, Lesher L, Schlievert PM, Rogers T, Villaume LG, Danila R, Lynfield R. 2011. Staphylococcal toxic shock syndrome 2000-2006: epidemiology, clinical features, and molecular characteristics. *PLoS One* 6:e22997.
4. Friedman ND. 1981. *Everything you must know about tampons*. Berkley Books, New York, NY.
5. Schlievert PM, Davis CC. Device-Associated Menstrual Toxic Shock Syndrome. Clin Microb Rev. 2020; 33(3).
6. Billon A, Gustin M, Tristan A, Benet T, Berthiller J, Gustave CA, Vanhems P, lina G. Association of characteristics of tampon use with menstrual toxic shock syndrome in France. EClinical Medicine. 2020; 21:100316.
7. DeVito MJ, Schecter A. Exposure Assessment to Dioxins From the Use of Tampons and Diapers. Enviro Health Pers. 2002; 110(1): 23-28.
8. Marcelis Q, Gatzios A, Deconinck E, Rogiers V, Desmedt B, Vanhaecke T. Quantitative risk assessment of allergens leaching from menstrual hygiene products. Regulatory Toxicology and Pharmacology. 2022; 135: 105260

9. DeLoughery E, Colwill AC, Edelman A, et al. Red blood cell capacity of modern menstrual products: considerations for assessing heavy menstrual bleeding. BMJ Sexual & Reproductive Health Published Online First: 07 August 2023.
10. Sun Y, Li C, Yan Y, Lai A, Peng X, Yue X, Li Y, Liu J, Liu Y. Effects of Hemp Sanitary Pads on the Vaginal Microecology. Comput Math Meth Med. 2022. Vol. 2022, Article ID 4435722.
11. Woeller KE, Hochwalt AE. Safety assessment of sanitary pads with a polymeric foam absorbent core. Regul Toxicol Pharmacol. 2015 Oct;73(1):419-24.
12. Government of Canada. Menstrual Tampons. 2015. <https://www.canada.ca/en/health-canada/services/drugs-medical-devices/menstrual-tampons.html>
13. Gurman E, How much do Canadian women ACTUALLY spend on tampons and pads each year? Chatelaine. 2017. <https://www.chatelaine.com/living/how-much-do-periods-cost/>
14. Manley H, Hunt JA, Santos L, Breedon P. Comparison between menstrual cups: first step to categorization and improved safety. Womens Health (Lond). 2021 Jan-Dec;17:17455065211058553.
15. Schnyer AN, Jensen JT, Edelman A, Han L. Do menstrual cups increase risk of IUD expulsion? A survey of self-reported IUD and menstrual hygiene product use in the United States. Eur J Contracept Reprod Health Care. 2019 Oct;24(5):368-372.
16. Long, Jill MD, MPH, MHS; Schreiber, Courtney MD, MPH; Creinin, Mitchell D. MD; Kaneshiro, Bliss MD, MPH; Nanda, Kavita MD, MPH; Blithe, Diana PhD. Menstrual Cup Use and Intrauterine Device Expulsion in a Copper Intrauterine Device Contraceptive Efficacy Trial [OP01-1B]. Obstetrics & Gynecology: May 2020 - Volume 135 - Issue - p 1S

CHAPTER 26. MENSTRUAL EQUITY AND ACTIVISM

1. Weiss-Wolf J. Periods Gone Public. 2019. Simon & Schuster.
2. Hennegan J, Winkler IT, Bobel C, Keiser D, Hampton J, Larsson G, Chandra-Mouli V, Plesons M, Mahon T. Menstrual health: a definition for policy, practice, and research.
3. Hensen B, Gondwe M, Phiri M, Schaap A, Simuyaba M, Floyd S, Mwenge L, Sigande L, Shanaube K, Simwinga M, Fiddler S, Hayes R, Ayles H. Access to menstrual hygiene products through incentivised, community-based, peer-led sexual and reproductive health services before and during the COVID-19 pandemic: findings from the Yathu Yathu trial. BMC Public Health. 2022 Mar 21;22(1):554.
4. Arshad Ali S, Baloch M, Riaz L, Iqbal A, Riaz R, Perveen B, Siddiqui M, Arshad Ali A. Perceptions, Practices, and Challenges Regarding Menstrual Hygiene Among Women in Karachi, Pakistan: A Comparison Between General Population and Healthcare Workers. Cureus. 2020 Aug 20;12(8):e9894.
5. Hennegan J, Shannon AK, Rubli J, Schwab KJ, Melendez-Torres GJ. Women's and girls' experiences of menstruation in low- and middle-income countries: A systematic review and qualitative metasynthesis. PLoS Med. 2019 May 16;16(5):e1002803.
6. van Eijk AM, Jayasinghe N, Zulaika G, Mason L, Sivakami M, Unger HW, Phillips-Howard PA. Exploring menstrual products: A systematic review and meta-analysis of reusable menstrual pads for public health internationally. PLoS One. 2021 Sep 24;16(9):e0257610.
7. Bozelko C. Opinion: Prisons that Withhold Menstrual Pads Humiliate Women and Violate Basic Rights. 2020 Jul 25. In: Bobel C, Winkler IT, Fahs B, Hasson KA, Kissling EA, Roberts TA, editors. The Palgrave Handbook of Critical Menstruation Studies [Internet]. Singapore: Palgrave Macmillan; 2020. Chapter 5.
8. Chowdhary P, Mekuria FT, Tewahido D, Gulema H, Derni R, Edmeades J. Building sustainable and scalable peer-based programming: promising approaches from TESFA in Ethiopia. Reprod Health. 2022 Jun 13;19(Suppl 1):55.
9. Schwartz BI, Effron A, Bear B, Short VL, Eisenberg J, Felleman S, Kazak AE. Experiences with Menses in Transgender and Gender Nonbinary Adolescents. J Pediatr Adolesc Gynecol. 2022 Aug;35(4):450-456.
10. Rossouw L, Ross H. Understanding Period Poverty: Socio-Economic Inequalities in Menstrual Hygiene Management in Eight Low- and Middle-Income Countries. Int J Environ Res Public Health. 2021 Mar 4;18(5):2571.

11. Parraga J. Period Poverty. Schulich School of Medicine & Dentistry, Western University. 2020. <https://www.schulich.uwo.ca/rapport/2020/topical/period_poverty.html>

CHAPTER 27: BRING IT BACK TO THE BASICS & CHAPTER 28. CYCLICAL LIVING

1. Daundasekara SS, Arlinghaus KR, Johnston CA. The Importance of Language in Behavior Change. Am J Lifestyle Med. 2019 Feb 19;13(3):239-242.
2. Sharpe M. Instagram Post, @marthasharpe_nd, April 22, 2024. <https://www.instagram.com/p/C6EfHJ5rPbA/>
3. Vitti A. Woman Code: Perfect Your Cycle, Amplify Your Fertility, Supercharge Your Sex Drive, and Become a Power Source. Harper Collins, Mar 19, 2013.
4. Statistics Canada. Canadian Health Measures Survey: Activity monitor data, 2018-2019. <https://www150.statcan.gc.ca/n1/daily-quotidien/210901/dq210901c-eng.htm>
5. Buckley T, Gottlieb A. A critical appraisal of theories of menstrual symbolism. In: Buckley T, Gottlieb A, editors. Blood magic. The anthropology of menstruation. Berkeley: University of California Press; 1988. pp. 3–50.
6. Del Mar Fernández M, Regueira-Méndez C, Takkouche B. Psychological factors and premenstrual syndrome: A Spanish case-control study. PLoS One. 2019 Mar 6;14(3):e0212557.
7. Mooney-Somers J, Perz J, Ussher JM. A complex negotiation: women's experiences of naming and not naming premenstrual distress in couple relationships. Women Health. 2008;47(3):57-77. doi: 10.1080/03630240802134134. Erratum in: Women Health. 2009 Mar-May;49(2-3):262. PMID: 18714712.

CHAPTER 29. NOURISHING YOUR CYCLE

1. Government of Canada. Canada's Food Guide. 2021. <https://food-guide.canada.ca/en/>
2. EduChange and NUPENS. NOVA Classification Reference Sheet. Food, Nutrition, & Fitness I: The Digestion Journey Begins with Food Choices. 2018, Sao Paulo.
3. Nikolopoulou, K. (2023, March 10). What Is Confirmation Bias? | Definition & Examples. Scribbr. Retrieved April 22, 2024, from https://www.scribbr.com/research-bias/confirmation-bias/
4. Orlich MJ, Singh PN, Sabaté J, Jaceldo-Siegl K, Fan J, Knutsen S, Beeson WL, Fraser GE. Vegetarian dietary patterns and mortality in Adventist Health Study 2. JAMA Intern Med. 2013 Jul 8;173(13):1230-8.
5. Segovia-Siapco G, Sabaté J. Health and sustainability outcomes of vegetarian dietary patterns: a revisit of the EPIC-Oxford and the Adventist Health Study-2 cohorts. Eur J Clin Nutr. 2019 Jul;72(Suppl 1):60-70.
6. Han MA, Zeraatkar D, Guyatt GH, Vernooij RWM, El Dib R, Zhang Y, Algarni A, Leung G, Storman D, Valli C, Rabassa M, Rehman N, Parvizian MK, Zworth M, Bartoszko JJ, Lopes LC, Sit D, Bala MM, Alonso-Coello P, Johnston BC. Reduction of Red and Processed Meat Intake and Cancer Mortality and Incidence: A Systematic Review and Meta-analysis of Cohort Studies. Ann Intern Med. 2019 Nov 19;171(10):711-720.
7. Ditano-Vázquez P, Torres-Peña JD, Galeano-Valle F, Pérez-Caballero AI, Demelo-Rodríguez P, Lopez-Miranda J, Katsiki N, Delgado-Lista J, Alvarez-Sala-Walther LA. The Fluid Aspect of the Mediterranean Diet in the Prevention and Management of Cardiovascular Disease and Diabetes: The Role of Polyphenol Content in Moderate Consumption of Wine and Olive Oil. Nutrients. 2019 Nov 19;11(11):2833.
8. Devranis P, Vassilopoulou E, Tsironis V, Sotiriadis PM, Chourdakis M, Aivaliotis M, Tsolaki M. Mediterranean Diet, Ketogenic Diet or MIND Diet for Aging Populations with Cognitive Decline: A Systematic Review. Life (Basel). 2023 Jan 6;13(1):173.
9. Djuricic I, Calder PC. Beneficial Outcomes of Omega-6 and Omega-3 Polyunsaturated Fatty Acids on Human Health: An Update for 2021. Nutrients. 2021 Jul 15;13(7):2421.
10. Papadaki A, Nolen-Doerr E, Mantzoros CS. The Effect of the Mediterranean Diet on Metabolic Health: A Systematic Review and Meta-Analysis of Controlled Trials in Adults. Nutrients. 2020 Oct 30;12(11):3342.

11. Brzezinski AA, Wurtman JJ, Wurtman RJ, Gleason R, Greenfield J, Nader T. d-Fenfluramine suppresses the increased calorie and carbohydrate intakes and improves the mood of women with premenstrual depression. Obstet Gynecol. 1990 Aug;76(2):296-301.
12. Thakur H, Pareek P, Sayyad MG, Otiv S. Association of Premenstrual Syndrome with Adiposity and Nutrient Intake Among Young Indian Women. Int J Womens Health. 2022 May 4;14:665-675.
13. Wei Y, Lv J, Guo Y, Bian Z, Gao M, Du H, Yang L, Chen Y, Zhang X, Wang T, Chen J, Chen Z, Yu C, Huo D, Li L; China Kadoorie Biobank Collaborative Group. Soy intake and breast cancer risk: a prospective study of 300,000 Chinese women and a dose-response meta-analysis. Eur J Epidemiol. 2020 Jun;35(6):567-578.
14. Takeda T, Chiba Y. Evaluation of a natural S-equol supplement in treating premenstrual symptoms and the effect of the gut microbiota: An open-label pilot study. Neuropsychopharmacol Rep. 2022 Jun;42(2):127-134.
15. Bryant M, Cassidy A, Hill C, Powell J, Talbot D, Dye L. Effect of consumption of soy isoflavones on behavioural, somatic and affective symptoms in women with premenstrual syndrome. Br J Nutr. 2005 May;93(5):731-9.
16. Reed KE, Camargo J, Hamilton-Reeves J, Kurzer M, Messina M. Neither soy nor isoflavone intake affects male reproductive hormones: An expanded and updated meta-analysis of clinical studies. Reprod Toxicol. 2021 Mar;100:60-67.
17. Li N, Wu X, Zhuang W, Xia L, Chen Y, Zhao R, Yi M, Wan Q, Du L, Zhou Y. Soy and Isoflavone Consumption and Multiple Health Outcomes: Umbrella Review of Systematic Reviews and Meta-Analyses of Observational Studies and Randomized Trials in Humans. Mol Nutr Food Res. 2020 Feb;64(4):e1900751.
18. Elliot A. A Mind Spread Out on the Ground. Anchor Canada: 2019.

CHAPTER 30. MINDSET AND LIFESTYLE

1. Witkoś J, Hartman-Petrycka M. The Influence of Running and Dancing on the Occurrence and Progression of Premenstrual Disorders. Int J Environ Res Public Health. 2021 Jul 27;18(15):7946.
2. Song C, Ikei H, Park BJ, Lee J, Kagawa T, Miyazaki Y. Psychological Benefits of Walking through Forest Areas. Int J Environ Res Public Health. 2018 Dec 10;15(12):2804. doi: 10.3390/ijerph15122804. Erratum in: Int J Environ Res Public Health. 2020 Feb 18;17(4).
3. Hersey, T. Rest is Resistance. Hachette Book Group, New York. 2022.
4. Song C, Ikei H, Kagawa T, Miyazaki Y. Effects of Walking in a Forest on Young Women. Int J Environ Res Public Health. 2019 Jan 15;16(2):229.
5. Wielgosz J, Goldberg SB, Kral TRA, Dunne JD, Davidson RJ. Mindfulness Meditation and Psychopathology. Annu Rev Clin Psychol. 2019 May 7;15:285-316. doi: 10.1146/annurev-clinpsy-021815-093423. Epub 2018 Dec 10.
6. Morin CM, Carrier J, Bastien C, Godbout R; Canadian Sleep and Circadian Network. Sleep and circadian rhythm in response to the COVID-19 pandemic. Can J Public Health. 2020 Oct;111(5):654-657. doi: 10.17269/s41997-020-00382-7. Epub 2020 Jul 22.
7. Burns AC, Saxena R, Vetter C, Phillips AJK, Lane JM, Cain SW. Time spent in outdoor light is associated with mood, sleep, and circadian rhythm-related outcomes: A cross-sectional and longitudinal study in over 400,000 UK Biobank participants. J Affect Disord. 2021 Dec 1;295:347-352. doi: 10.1016/j.jad.2021.08.056. Epub 2021 Aug 27.
8. Oliviero F, Marmugi A, Viguié C, Gayrard V, Picard-Hagen N, Mselli-Lakhal L. Are BPA Substitutes as Obesogenic as BPA? Int J Mol Sci. 2022 Apr 11;23(8):4238.
9. Werkneh AA, Gebru SB, Redae GH, Tsige AG. Removal of endocrine disrupters from the contaminated environment: public health concerns, treatment strategies and future perspectives - A review. Heliyon. 2022 Mar 31;8(4):e09206.
10. Virant-Klun I, Imamovic-Kumalic S, Pinter B. From Oxidative Stress to Male Infertility: Review of the Associations of Endocrine-Disrupting Chemicals (Bisphenols, Phthalates, and Parabens) with Human Semen Quality. Antioxidants (Basel). 2022 Aug 20;11(8):1617.
11. Segovia-Mendoza M, Palacios-Arreola MI, Monroy-Escamilla LM, Soto-Piña AE, Nava-Castro KE, Becerril-Alarcón Y, Camacho-Beiza R, Aguirre-Quezada DE, Cardoso-Peña E, Amador-Muñoz O,

Garduño-García JJ, Morales-Montor J. Association of Serum Levels of Plasticizers Compounds, Phthalates and Bisphenols, in Patients and Survivors of Breast Cancer: A Real Connection? Int J Environ Res Public Health. 2022 Jun 30;19(13):8040.

12. National Institute of Environmental Health Sciences. "Environmental Agents." Updated 2022. <https://www.niehs.nih.gov/health/topics/agents/index.cfm>
13. Fonseca BM, Rebelo I. Cannabis and Cannabinoids in Reproduction and Fertility: Where We Stand. Reprod Sci. 2022 Sep;29(9):2429-2439. doi: 10.1007/s43032-021-00588-1. Epub 2021 May 10.

CHAPTER 31. ADVOCATING FOR YOUR CYCLE

1. Schatz AA, Brooks-Coley K, Harrington E, Murray MS, Carlson RW. Patient, Caregiver, and Oncologist Experiences With and Perceptions of Racial Bias and Discrimination in Cancer Care Delivery. J Natl Compr Canc Netw. 2022 Oct;20(10):1092-1098.
2. Weisz G, Knaapen L. Diagnosing and treating premenstrual syndrome in five western nations. Soc Sci Med. 2009 Apr;68(8):1498-505.
3. Soled KRS, Dimant OE, Tanguay J, Mukerjee R, Poteat T. Interdisciplinary clinicians' attitudes, challenges, and success strategies in providing care to transgender people: a qualitative descriptive study. BMC Health Serv Res. 2022 Sep 8;22(1):1134.
4. Zelin NS, Scott C, Avila-Quintero VJ, Curlin K, Flores JM, Bloch MH. Sexual Orientation and Racial Bias in Relation to Medical Specialty. J Homosex. 2022 Oct 21:1-26.
5. Ridd M, Shaw A, Lewis G, Salisbury C. The patient-doctor relationship: a synthesis of the qualitative literature on patients' perspectives. Br J Gen Pract. 2009 Apr;59(561):e116-33.
6. Daundasekara SS, Arlinghaus KR, Johnston CA. The Importance of Language in Behavior Change. Am J Lifestyle Med. 2019 Feb 19;13(3):239-242.

COMING 2026

The PCOS Literacy Handbook

Made in the USA
Monee, IL
13 September 2026

62981955R10221